SPACE CRAFT

Brimming with creative inspiration, how-to projects, and useful information to enrich your everyday life, Quarto Knows is a favorite destination for those pursuing their interests and passions. Visit our site and dig deeper with our books into your area of interest: Quarto Creates, Quarto Cooks, Quarto Homes, Quarto Lives, Quarto Drives, Quarto Explores, Quarto Gifts, or Quarto Kids.

Inspiring | Educating | Creating | Entertaining

© 2018 Quarto Publishing Group USA Inc.
Text © 2018 Michael Gorn
Illustrations © 2018 Giuseppe de Chiara

First published in 2018 by Voyageur Press, an imprint of The Quarto Group, 401 Second Avenue North, Suite 310, Minneapolis, MN 55401 USA. Telephone: (612) 344-8100 Fax: (612) 344-8692

www.QuartoKnows.com

Voyageur Press titles are also available at discount for retail, wholesale, promotional, and bulk purchase. For details, contact the Special Sales Manager by email at specialsales@quarto.com or by mail at The Quarto Group, Attn: Special Sales Manager, 401 Second Avenue North, Suite 310, Minneapolis, MN 55401 USA.

10 9 8 7 6 5 4 3 2 1

ISBN: 978-0-7603-5418-6

Library of Congress Cataloging-in-Publication Data

Names: Gorn, Michael H., author.
Title: Spacecraft : 100 iconic rockets, shuttles, and satellites that put us into space / by Michael Gorn.
Description: Minneapolis, Minnesota : Voyageur Press, 2018. | Includes bibliographical references and index.
Identifiers: LCCN 2018002624 | ISBN 9780760354186 (hc)
Subjects: LCSH: Launch vehicles (Astronautics)--History.
Classification: LCC TL785.8.L3 .G67 2018 | DDC 629.409--dc23
LC record available at https://lccn.loc.gov/2018002624

Acquiring Editor: Dennis Pernu
Project Manager: Jordan Wiklund
Art Direction and Cover Design: Cindy Samargia Laun
Page Design and Layout: Simon Larkin

Printed in China

MIX
Paper from responsible sources
FSC® C008047

On the front cover: The main boosters of Space Shuttle Atlantis light up the launch pad as the shuttle rockets towards space. *NASA Image Archive*
On the back cover: On a clear day, Space Shuttle Atlantis achieves liftoff once again. *Shutterstock*
On the endpapers: A collage of spacecraft by Giuseppe De Chiara.

PHOTO CREDITS
NASA Image Archive: 6, 7, 8, 9, 10, 12, 23, 45, 65, 114 (top and bottom), 115, 122, 144, 145, 172, 173 (top and bottom), 174, 188
Shutterstock: 94, 104, 125, 133, 155, 186
S. P. Korolev Rocket & Space Corporation (ENERGIA): 7, 48, 98, 102

SPACE CRAFT

100 ICONIC ROCKETS, SHUTTLES, AND SATELLITES THAT PUT US IN SPACE

Giuseppe De Chiara and Michael H. Gorn

VOYAGEUR PRESS

CONTENTS

Introduction 6
Selected sources
    and further reading 220
Index 222
Acknowledgments 224

1

THE FIRST SPACE AGE
1957-1977

CAPSULES

Mercury 12
Gemini 15
Apollo 19
Vostok 3K 24
Voskhod 3KV 27
Voskhod 3KD 28
Soyuz 7K-OK 30
Soyuz 7K-OKS 33
Soyuz 7K-L1 34
Soyuz 7K-LOK 34
Soyuz 7K-T 37
Soyuz 7K-TM (ASTP) 37
Progress 7K-TG 38

SPACEPLANES

X-15 38
MiG 105-11 40

LANDERS

LK 42
Lunokhod Rovers 1 and 2 43

STATIONS

Skylab 44
Apollo-Soyuz Test Project 47
Salyut 1 53
Salyut 2, 3, 5 (Almaz) 54
Salyut 4 54

ROCKETS

Vanguard 1 56
Redstone/Juno-1 59
Mercury-Atlas 59
Agena 60
Centaur 62
Gemini-Titan II 63
Saturn IB 66
Saturn V 69
R7 70
N1 73
UR-500 Proton 73

ROBOTICS

Explorer 1 75
Vanguard 1 76
Lunar Orbiter 1 78
Ranger 7 81
Surveyor 1 81
Mariner 10 82
Pioneer 10 85
Viking 1 and 2 86
Voyager 2 88
Sputnik 1 91
Sputnik 2 93
Sputnik 3 94
Luna 3 97
Luna 4-9 97
Luna 15, 16, 18, 20 98
Venera 9-12 101

2

THE SECOND SPACE AGE
1977-1997

CAPSULES
Soyuz 7K-ST 106
Soyuz-TM (7K-STM) 110
TKS 110

SPACEPLANES
Space Transportation System 111
The Shuttle Orbiters 114
Orbiter Enterprise 121
VKK Buran 126

STATIONS
Mir 129
Atlantis-Mir 129
Salyut 6 and 7 132

ROCKETS
Shuttle Rocketry 135
Ariane 4 137
Energia 137

ROBOTICS
Cassini-Huygens 139
Hubble Space Telescope 141
Compton Gamma Ray Observatory ... 146
Galileo 149
Giotto 150
Ulysses 152

3

SPACE EXPLORATION AT A CROSSROADS
1997-2017

CAPSULES
Orion 156
Dragon 159
CST-100 160
Shenzhou 161
Soyuz MS (7K-MS) 163

SPACEPLANES
SpaceShipOne and Two 164
X-37B 167

STATIONS
International Space Station 168
Unity 174
Zarya 175
Zvezda 176
Destiny Laboratory 178
Multipurpose Logistics Modules 178
Tiangong-1 180

ROCKETS
Space Launch System 182
Falcon 9 182
Antares 189
New Shepard 190
Ariane 5 193
VEGA 194
Long March-2F 194

ROBOTICS
Opportunity and Curiosity 197
Chandra Telescope 200
Spitzer Space Telescope 203
Juno 204
Deep Impact 207
New Horizons 208
James Webb Space Telescope 208
BepiColombo 211
Herschel and Planck 212
LISA Pathfinder 216
Rosetta 219

INTRODUCTION

Spacecraft tells the story of a momentous period in world history, when we escaped the atmosphere, looked down at the seas and the continents, and for the first time saw the planet in all its complexity, fragility, and beauty. At the same time, we looked skyward into the heavens and saw celestial phenomena as old as creation and as transcendent as the mind can fathom. The daring vehicles that accomplished these missions during the last six decades—from Sputnik 1 to the James Webb telescope—represent the summation of space age exploration up to the early twenty-first century; they reflect the ongoing attempt by humanity to comprehend the workings of the universe and its parts.

The initial forays into space had Earth-bound parallels. At the dawn of the Space Age in the 1950s, President Dwight D. Eisenhower proposed legislation for a uniform network of high-speed freeways across the United States, ostensibly for national defense. Its usefulness, however, proved far wider. The National Interstate and Defense Highways Act of 1956 represented the biggest public works expenditure in US history (about $41 billion by the end of the Eisenhower administration alone) and expanded the American expressway grid by roughly 41,000 miles. More than that, it stimulated job creation and contributed to the economic boom of the 1950s, accelerated the pace of modern commerce, and narrowed sectional and regional differences, geopolitically unifying the country as never before. In their own way, the spacecraft that serve humanity return many of the same benefits.

Three years before the passage of this law, on the other side of the world, a New Zealander and a Nepalese Sherpa undertook a completely unrelated challenge. On May 29, 1953, Edmund P. Hillary and Tenzing Norgay became the first people to reach the summit of Mount Everest, the tallest point on Earth at 29,028 feet (8,848 meters). "We didn't know if it was humanly possible to reach the top of Mount Everest," Hillary later said. Of no military, economic, or political consequence, reaching the peak of the world satisfied several deep-seated human drives: the pursuit of origins, the fulfillment of curiosity, the lure of adventure, the revelation of the unknown, and the test of endurance. These same motivations also typify the urge to explore space.

Because this book is confined to spacecraft designed for illumination rather than for tangible results—for endeavors like Mount Everest, rather than the National Highway System—there arises the inevitable debate between human and robotic sojourners. Will we engage with the universe through cosmonauts, astronauts, and taikonauts—through sentient explorers—or will automated orbiters and landers be our surrogates? The spacecraft profiled suggest that both have unique but complementary roles, that robots will continue to be sent to places too inhospitable or too distant for us, and that people will go where risk can be managed and where human intelligence cannot be substituted.

The Mars Science Laboratory's Curiosity rover performs many of the same chores as living spacefarers, without the limiting factors of time, fuel, supplies, and (to name only one safety issue) radiation exposure—the latter a serious health concern according to Curiosity's own findings. On the other hand, the twelve Apollo astronauts who walked on the moon from 1969 to 1972 accomplished tasks impossible for machines. They witnessed and described the palpable texture of the moon's surface, assessed the general features of its geology, and felt the peculiarities of lunar gravity in their every step and motion. They quarried large and highly varied quantities of rock and soil (842 pounds in all). They reported changes in their own physiology and psychology. They golfed! Then they returned to Earth with suggestions for future missions and explained how the experience affected them in artwork, memoirs, and the media.

Spacecraft also illustrates the international context of space travel over the last sixty years. It started as a duel between the United States and USSR, in the height of the Cold War, that not only involved a technological contest, but a struggle for world opinion. The Soviets and Americans trumpeted their space achievements to the world as proof of the superiority of their cultures, their governments, their economic systems, and their political ideologies. But once the Cold War ended, space exploration ceased to be a subplot in a global confrontation and in time assumed a radically different character.

This evolution underlies the organization of this book. Divided into three parts of twenty years each, it traces three distinct periods. The initial one, from 1957 to 1977, covers the first Space Age, during which time the two superpowers vied for supremacy. The climax occurred with the race to the moon, a battle won by the United States in part because President John F. Kennedy imposed the daring objective of a lunar landing. That accomplishment erased the early lead held by the Soviets and enabled NASA to leap ahead to the finish line. But even though Apollo captured the big prize, the clash did not end there. To redeem themselves, the Soviets pivoted to new ground. By pursuing space station development with the seven Salyuts and Mir, the USSR restored its reputation and closed out the first phase of space exploration with an unquestioned mastery of long-duration habitation in orbit. The United States, meanwhile, scored its own late-innings win by concentrating on a series of spectacular robotic missions: the historic flights of Vikings 1 and 2 to Mars in the bicentennial year and the so-called Grand Tour of Voyagers 1 and 2 to the outer planets starting in 1977.

The next point on the timeline—from 1977 to 1997—was a transition period that left behind the Soviet-American polarity and opened a new era of tentative, multinational collaborations. This realignment began with the recognition by NASA and the Russian space agency that the end of the Cold War and the tightening of budgets left neither side with the resources necessary to build massive new projects in space. Consequently, to the shock of many, these former combatants agreed to forget (if not forgive) their old hostilities and to negotiate a partnership for an immensely

ambitious undertaking: the International Space Station (ISS). Along with the ISS, the space shuttle also became a key contributor to the budding détente in the heavens. Its large crew complement of up to seven enabled travelers of many nations, races, and creeds to work and live beside one another, and its cavernous cargo hold rendered it the sole means of transporting and assembling the outsized modules of the ISS. This era also marked the emergence of another transnational force: the European Space Agency (ESA), whose ten founding states came to epitomize fully integrated teamwork across borders.

Finally, during the third phase—entitled Space Exploration at a Crossroads, 1997 to 2017—the study of the universe and its constituents became a global initiative. The docking of the ISS's first two modules in 1998 signified the realization of a massive multinational project with a lifespan expected to last at least until 2024, under the supervision of a five-member consortium consisting of the United States, Russia, Canada, Japan, and the ESA. At the same time, ESA's own program gained in stature, producing some of the world's most sophisticated robotic spacecraft and launch vehicles. It also expanded its roster to include twenty-two member countries, embracing most of the European continent (ESA's governing council consists of Austria, Belgium, the Czech Republic, Denmark, Estonia, Finland, France, Germany, Greece, Hungary, Ireland, Italy, Luxembourg, the Netherlands, Norway, Poland, Portugal, Romania, Spain, Sweden, Switzerland, and the United Kingdom).

The Russian Federal Space Agency continued to pursue advanced and reliable rocketry (expected to carry American astronauts to and from the ISS for as many as ten years after the shuttle's retirement in 2011) and produced improved versions of the time-honored Soyuz transports. Meantime, in the United States, NASA awarded major contracts to several small, private firms, such as SpaceX, Blue Origins, and Orbital ATK, to fabricate capsules and rockets capable of sending astronauts into space. Doing so presented an unprecedented challenge to one of the main orthodoxies of the space age: that only central governments reserved the right to engage in missions of discovery. And perhaps more decisive than any other recent development, the China National Space Administration became only the third country to launch its own citizens into orbit, achieving a high rate of success and making ambitious plans for the future.

Despite this trend toward broad-based partnerships and a more crowded field of participants, a troubling question loomed over twenty-first century space exploration: Will the present pattern prevail, or will a rising tide of populism and nationalism incite a return to the early space age, when political adversaries projected their earthly antagonisms into the cosmos?

1

THE
FIRST
SPACE
AGE

1957–1977

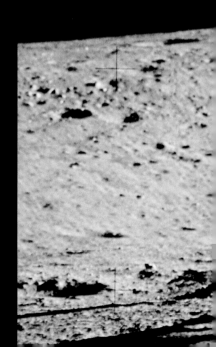

CAPSULES
Mercury

NASA opened its doors on October 1, 1958, and the following month Project Mercury—named for the winged messenger of the gods in Roman mythology—got its start as the agency's highest priority project. Before awarding the prime contract to McDonnell Douglas in 1959, the Space Task Group defined the capsule's specifications: 6.8 feet (2 meters) long, 6.8 feet (1.9 meters) in diameter (at its base), and up to 3,000 pounds (1,400 kilograms) at launch. Stuffed to the limit with equipment, the interior furnished the sole astronaut with a tight, foam-fitted seat and just 36 cubic feet (1 cubic meter) of habitable volume— about the size of a two-cushion love seat. The capsule's small size and low weight reflected the relatively weak thrust of its launch vehicles.

The Mercury spacecraft consisted of six main structural components: an escape tower for crew egress; an antenna housing at the top of the capsule; reentry parachutes in the narrow neck of the spacecraft; the crew compartment in the broad lower section; a heat shield below it; and the retrorockets at the very bottom. As initially conceived, the Mercury spacecraft's flight path could only be guided from the ground. But in due course, the astronauts persuaded the designers to build cockpit controls, which in a few cases proved to be lifesaving.

Mercury experienced an unpredictable mixture of failure and success during the flight tests that preceded missions with astronauts on board. The first launch, in September 1959—a research and development mission—went aloft on an Atlas D missile and flew according to plan. But in July 1960, a ballistic-trajectory test of the Atlas missile/capsule combination ended in failure. The following November, the first of the numbered Mercury flights, known as Mercury-Redstone 1, also miscarried when its rocket engines shut down just after liftoff. The program recovered with a second attempt at Mercury-Redstone 1 in December 1960. The next month, NASA attempted to send a living creature into space—not an astronaut, but a chimpanzee named Ham, who flew on Mercury-Redstone 2. Ham survived some serious scares, enduring 17 gs at liftoff, an off-target splashdown, and a leaking spacecraft.

Mercury-Redstones 3 and 4 finally launched Americans beyond the confines of Earth, but not before Yuri Gagarin became the first man in space on April 12, 1961, on board a Vostok 3K spacecraft. The United States found itself behind the USSR after the launch of Sputnik I and three and a half years later still had much ground to gain. Gagarin had *orbited* the Earth for 108

minutes; Alan Shepard, who flew on May 5, 1961, aboard Mercury-Redstone 3 (better known as Freedom 7), made a suborbital leap into space that lasted a little more than fifteen minutes. His mission went off without incident, but that of Virgil "Gus" Grissom on July 21, 1961, on board Mercury-Redstone-4 (Liberty Bell 7) did not fare as well. After a similar fifteen-minute journey, he splashed down, but the spacecraft's new explosive hatch cover unexpectedly blew open, filling the capsule with water and sinking it, leaving Grissom to swim to the surface, where a helicopter rescued him.

Despite the checkered record of the early Atlas launch vehicles, on February 20, 1962, NASA—under intense pressure to respond to Yuri Gagarin's flight ten months earlier—took a calculated gamble and began the countdown on American orbital spaceflight. From the pool of applicants submitted in 1958, Robert Gilruth and the Space Task Group at NASA Langley chose fourteen for Astronaut Group 1, and from this number selected seven principals for Mercury. Among them was Marine pilot (later colonel) John H. Glenn for the first orbital flight. He served as a combat pilot during World War II and the Korean conflict, after which he attended the US Navy test pilot school.

Glenn went into orbit on Friendship 7/Mercury-Atlas 6 on February 20, 1962, and he accomplished his mission, but not

McDonnell Mercury Spacecraft Design Evolution

0   1   2
meters

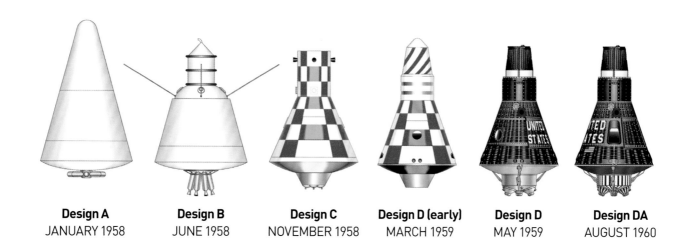

Design A
JANUARY 1958

Design B
JUNE 1958

Design C
NOVEMBER 1958

Design D (early)
MARCH 1959

Design D
MAY 1959

Design DA
AUGUST 1960

McDonnell Mercury
Design D-1 "MR-3 Flight Configuration"

0   1   2
meters

FRONT VIEW

TOP VIEW

SIDE VIEW

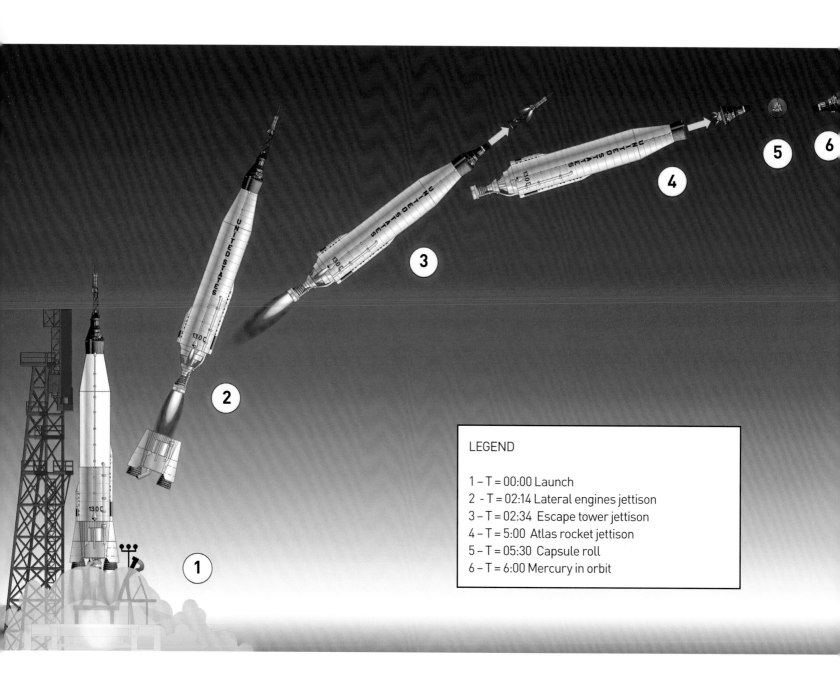

LEGEND

1 – T = 00:00 Launch
2 – T = 02:14 Lateral engines jettison
3 – T = 02:34 Escape tower jettison
4 – T = 5:00 Atlas rocket jettison
5 – T = 05:30 Capsule roll
6 – T = 6:00 Mercury in orbit

without some nail-biting moments. Flying over Mexico, in his second orbit, he noticed a loss of attitude control; the automatic stabilization system failed to keep the capsule on course, causing it to drift to the right about 1.5 degrees per second. To compensate, for the rest of the flight he flew the spacecraft manually (proving the value of trained pilots in the cockpit). But as he crossed over Cape Canaveral, an equally dangerous problem threatened the mission: the heat shield had become unlocked, according to sensors. Unsure whether this reflected a crisis or a false alarm, Glenn received instructions to ignore procedure and refrain from jettisoning the retropackage after he fired the retrorockets during reentry, in the hope that the retropackage's straps would hold the shield in place. As he descended into the atmosphere, everyone in ground control held their breath, but the mission ended safely after three orbits and nearly five hours in space. Glenn got lucky; the sensor had given an incorrect reading and the heat shield remained in place throughout the crisis.

More tense episodes remained before Mercury ended. In May 1962, Scott Carpenter in Aurora 7/Mercury-Atlas 7 made a flight of similar duration to Glenn's, but his capsule overshot the landing target by 261 miles (420 kilometers). Rescuers found Carpenter almost an hour after splashdown, sitting in a life raft after escaping from the capsule. Walter Schirra flew next aboard Sigma 7/Mercury-Atlas 8 in October 1962 and doubled the times logged in space by Glenn and Carpenter, making six orbits in more than nine hours with no incidents. At last, Gordon Cooper in Faith 7 lifted off atop Mercury-Atlas 9 in May 1963. This mission vastly expanded the range (and prestige) of the US space program. He remained in orbit for twenty-two revolutions, timed at thirty-four hours and twenty minutes. But some major troubles occurred. On the nineteenth orbit, an instrument light warned of deceleration and premature reentry; on the twenty-first, the automatic stabilization and system short-circuited and the carbon dioxide level rose in the spacecraft. Cooper took the controls, and, like Glenn, he flew his spacecraft to a safe reentry and splashdown.

Gemini

Although the early US space program seems, in hindsight, to run straight through from Projects Mercury to Gemini without interruption, in fact they had completely different origins and purposes. Mercury sprang from NASA's predecessor, the National Advisory Committee for Aeronautics; Gemini originated with a speech made by President John F. Kennedy to Congress on May 25, 1961. Kennedy's address came in response to when Soviet cosmonaut Yuri Gagarin became the first person in space on April, 1961, flying for 108 minutes in one orbit around the Earth. In sharp contrast, astronaut Alan Shepard made a short suborbital flight on May 5 that lasted slightly more than 15 minutes. The juxtaposition of these two space flights, separated by only twenty-two days, made it clear that the US remained far behind the USSR in space exploration, a serious deficiency at a time when the two superpowers vied for Cold War predominance.

The president asked the House and Senate for money (lots of money; he had been told that in the end it would cost about $33 billion) to land astronauts on the moon by the close of the 1960s. Congress approved, but the engineers and managers on Mercury realized that even if their project fulfilled its objectives, JFK's towering new goal demanded that they overcome a big technological deficit. For one thing, although Mercury had dispelled many of the fears about the effects of space travel on human beings, it did not yield sufficient physiological information for prolonged moon shots. Additionally, no Mercury flight attempted extravehicular activities, or EVAs, essential to master before astronauts walked on the moon. Even more important, the rendezvous and docking of spacecraft, without which no lunar program could occur, had yet to be tried or tested. But most fundamental of all, no consensus existed in 1961 among NASA's leaders about the overall architecture of the moon missions—how to send the astronauts there, how to land them, and how to bring them home.

Robert Gilruth and his Space Task Group at NASA Langley responded to Kennedy's commitment by conceiving of Project Gemini, named for the mythological Greek twins Castor and Pollux, the guardians of mariners. This duality reflected the two-man crews planned for the Gemini missions. To accelerate Gemini's development, NASA again turned to the US Air Force, not only choosing its Titan II missile as the Gemini launch vehicle, but also selecting the versatile Agena second stage as a rendezvous and docking partner for the Gemini capsule.

Two years passed from the last Mercury to the first Gemini flight, which seemed like an eternity to those in NASA responsible for closing the gap with the USSR. But the wait occurred because Gemini represented much more than an intermediate step on the way to the moon; rather, it embodied a technological leap over Mercury. Among its many advances, Gemini carried a 50-pound (22.7-kilogram) computer, the first in space, and capable of seven thousand calculations per second. The capsule came equipped with fuel cells that generated electrical power through the chemical reaction of oxygen and hydrogen. The process produced drinking water as a byproduct.

The capsule itself surpassed that of Mercury in every sense. Gemini weighed 8,360 pounds (3,792 kilograms) at launch, more than twice that of Mercury. Its length of 18.3 feet (5.60

meters) exceeded Mercury's by two and a half times, and its base diameter of 10 feet (3.05 meters) measured almost 4 feet wider than Mercury's. In the starkest and most meaningful contrast, Mercury contained just 36 cubic feet (1 cubic meter) of habitable space versus Gemini's 90 cubic feet (2.55 cubic meters). Even doubling the crew size to two still left the Gemini astronauts with a far more comfortable and effective cabin than Mercury's.

Gemini swung wildly between big achievements and hair-raising suspense. After a tranquil, three-orbit Gemini 3 in March 1965, Gemini 4 followed in June of that year with the failure of its much-anticipated on-board computer, leaving James McDivitt to fly the reentry manually. Not only that, but after Ed White made the first American EVA, he and McDivitt faced calamity as they exhausted themselves before finally closing and latching the spacecraft's faulty egress hatch. In August 1965, Gemini 5 lasted for eight days, giving hope that astronauts could survive in space for the full length of a round-trip lunar voyage. In March 1966, disaster almost struck again when Apollo 8 Commander Neil Armstrong and Pilot David Scott became the first to dock with another vehicle in space, joining with an Agena, but a jammed Gemini thruster caused the linked vehicles to tumble and spin,

and the astronauts survived only by firing their retrorockets and returning home two days early. Just three months later, Gemini 9's docking mission failed when the Agena did not achieve orbit. Gemini 10 in July 1966 restored some confidence when Michael Collins and John Young docked with an Agena, fired its engine, and boosted the conjoined spacecraft into a higher orbit, later rendezvousing with the still orbiting Gemini 8 Agena. Gemini 11 in September 1966 made a direct-ascent rendezvous with the Agena, and in a forty-four-minute spacewalk, Richard Gordon connected the capsule and the target vehicle with a 98.4-foot (30-meter) tether. This mission also witnessed the first fully computer-controlled reentry. Finally, in November 1966, Gemini 12 sent James Lovell and Buzz Aldrin into orbit, during which Aldrin made three successful EVAs lasting a combined five and a half hours.

Even though Gemini contributed enormously to the later moon expeditions, at its end only a little more than three years remained before the deadline on President Kennedy's lunar landing. And as the curtain closed on Gemini, no one in the space agency really knew whether astronauts would walk on the moon before the clock ran out.

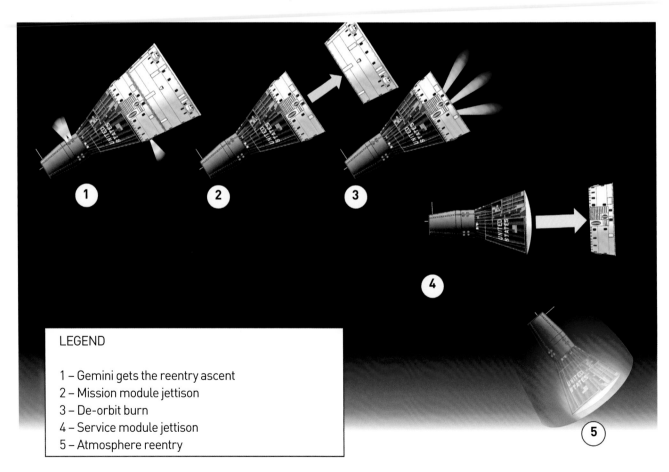

LEGEND

1 – Gemini gets the reentry ascent
2 – Mission module jettison
3 – De-orbit burn
4 – Service module jettison
5 – Atmosphere reentry

McDonnell Manned Spacecraft Design Evolution

0 1 2 meters

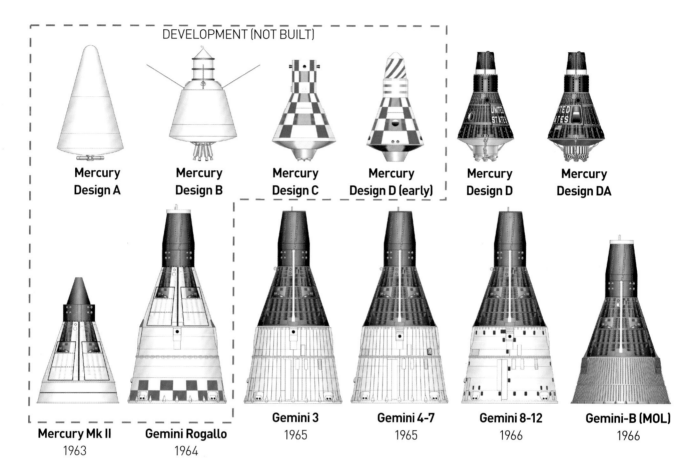

DEVELOPMENT (NOT BUILT)

**Mercury
Design A**

**Mercury
Design B**

**Mercury
Design C**

**Mercury
Design D (early)**

**Mercury
Design D**

**Mercury
Design DA**

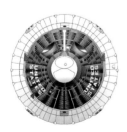

Mercury Mk II
1963

Gemini Rogallo
1964

Gemini 3
1965

Gemini 4-7
1965

Gemini 8-12
1966

Gemini-B (MOL)
1966

McDonnell Gemini SC 3
Launch Configuration

0 1 2 meters

FRONT VIEW

TOP VIEW

SIDE VIEW

North American Apollo Missions
Block 2 (H Missions)

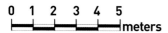
meters

TOP VIEW

FRONT VIEW

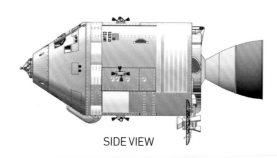
SIDE VIEW

North American Apollo Spacecraft
Block 2

0 1 2 3 4 5 meters

FRONT SECTION

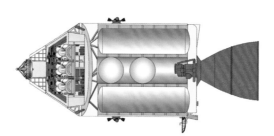
TOP SECTION

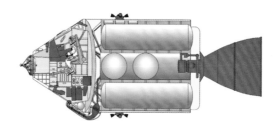
SIDE SECTION

Apollo

The Apollo program resulted from political desperation. Just three months after his inauguration, President John F. Kennedy found his administration under siege by two blows to American leadership and pride. The first came on April 12, 1961, when Soviet cosmonaut Yuri Gagarin orbited the Earth, becoming the first human to be launched into space. The second happened less than a week later when Cuban insurgents trained by the CIA went ashore at the Bay of Pigs on April 17, 1961, only to be routed by the forces of Fidel Castro.

Kennedy needed a bold action that would take the Soviets by surprise. On May 25, 1961, he addressed Congress in a speech and requested massive funding for a lunar landing before the end of the decade.

Congress voted for it, and NASA soon embarked on Project Gemini as a halfway step toward JFK's extravagant goal. Kennedy's speech came as a thunderbolt even to some of the highest-ranking space agency officials—almost none of whom had any inkling of the decision beforehand. As the shock sank in, these engineers and scientists realized that no rocket and no spacecraft existed that could even begin to fulfill what the president promised. But they did agree on a name for the project. They followed the mythological trend of Mercury and Gemini by naming the moon program in honor of Apollo, the son of Zeus, the greatest of the Greek gods.

Before any serious work could begin on Apollo, NASA needed to plot out the fundamental architecture for a moon shot. During 1961 and 1962, three main concepts competed aggressively for acceptance. The first, proposed by Dr. Wernher von Braun's team at NASA's Marshall Space Flight Center in Huntsville, Alabama, became known as **direct ascent**. It required a massive, as yet undeveloped rocket called the Nova that would carry crews straight to the moon, land, and return them to Earth. Direct ascent had powerful supporters, but its detractors dismissed it because of technical obstacles, expense, and lengthy development time. In contrast, the **earth-orbit rendezvous** technique envisioned the use of rockets to launch lunar modules into orbit, where they would be assembled before the voyage to the moon. This plan also won many backers (eventually von Braun himself) because it offered fewer technical challenges and a lower cost than direct ascent.

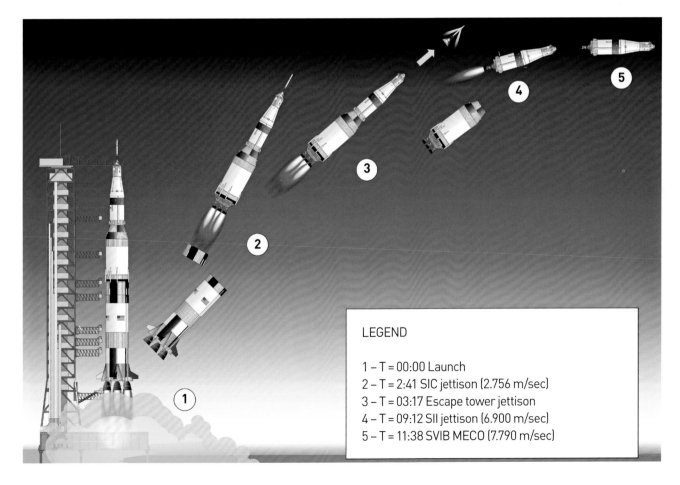

LEGEND

1 – T = 00:00 Launch
2 – T = 2:41 SIC jettison (2.756 m/sec)
3 – T = 03:17 Escape tower jettison
4 – T = 09:12 SII jettison (6.900 m/sec)
5 – T = 11:38 SVIB MECO (7.790 m/sec)

The final candidate encountered the most resistance. Called **lunar-orbit rendezvous** (LOR), it originated with a relatively obscure structures scientist at NASA Langley, John C. Houbolt. Houbolt's plan involved a spacecraft that did not require construction in space because it flew into orbit with three, already conjoined sections: a command module, a service module, and a lunar lander. It also required just one rocket, the Saturn V—a less powerful booster than the direct ascent's Nova, but strong enough to lift the multipart spacecraft out of the atmosphere prior to its trip to the moon. Once it reached its destination, the command module orbited the moon continuously while the lunar module detached itself, flew to the surface, landed, and undertook its mission. Upon completion, it launched itself from the surface and returned to dock with the circling command module for the return home.

Houbolt's LOR enjoyed a significant advantage over the other two options in that it accomplished the objective with lower cost, less fuel, and reduced weight. But it also entailed a serious liability in that the most error-prone parts of the trip—the lunar module's landing and takeoff from the moon, and its subsequent docking with the command module—occurred 240,000 miles (386,000 kilometers) from Earth, too far away for any practical rescue if things went awry.

Houbolt spent much of 1961 trying to persuade NASA's leaders of the validity of his system, most of whom greeted it with skepticism at best. But in the end, his exacting calculations persuaded von Braun and the Marshall Space Flight Center to endorse LOR in June 1962, after which others began to fall into line. The following month, Administrator James Webb backed LOR.

With these debates settled, the space agency plunged into the technical challenges of Apollo. North American Aviation won the prime contract for the command and service module, which took seven years to complete. Its design team soon appreciated the complexity of John Houbolt's LOR concept. The command module needed to orbit the Earth, fly to the moon, orbit it, and return home for a landing in the sea; sustain its three passengers for two weeks; and function seamlessly with its connected service module. The service module needed to store supplies, life-support equipment, fuel, maneuvering rockets, oxygen, and to house the all-important reentry engine.

Apollo CSM Family Development

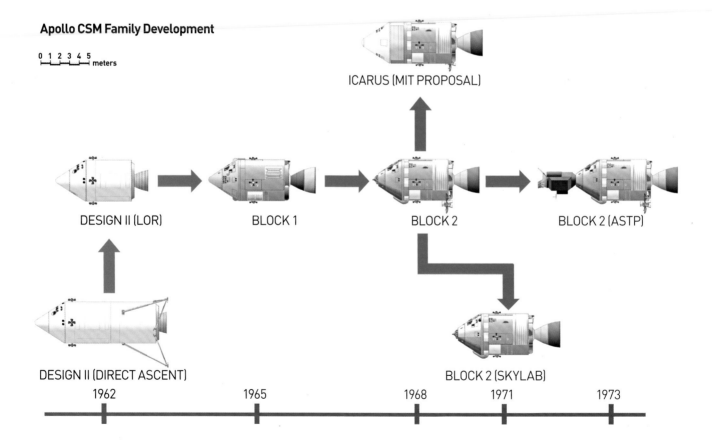

0 1 2 3 4 5 meters

ICARUS (MIT PROPOSAL)

DESIGN II (LOR)   BLOCK 1   BLOCK 2   BLOCK 2 (ASTP)

DESIGN II (DIRECT ASCENT)   BLOCK 2 (SKYLAB)

1962   1965   1968   1971   1973

Grumman LM – H Series

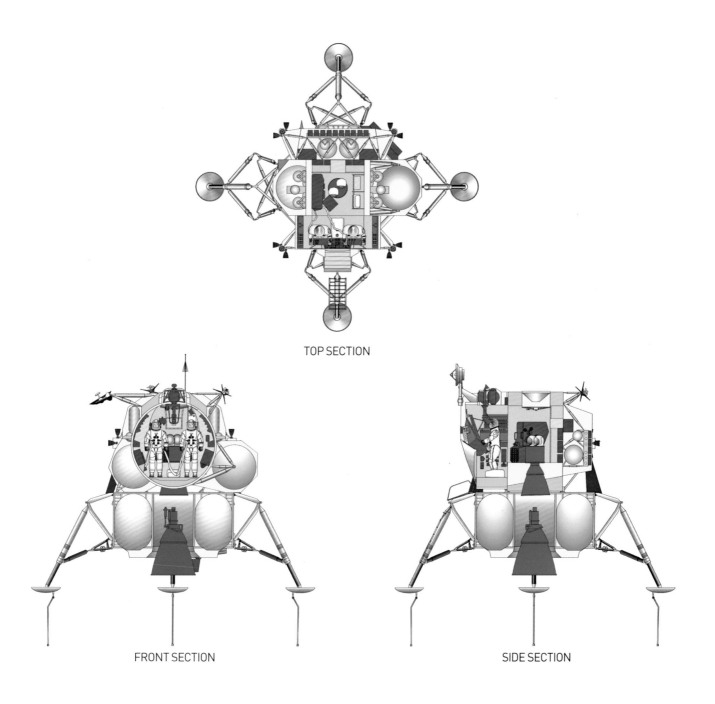

TOP SECTION

FRONT SECTION

SIDE SECTION

0 1 2 3 4 5
meters

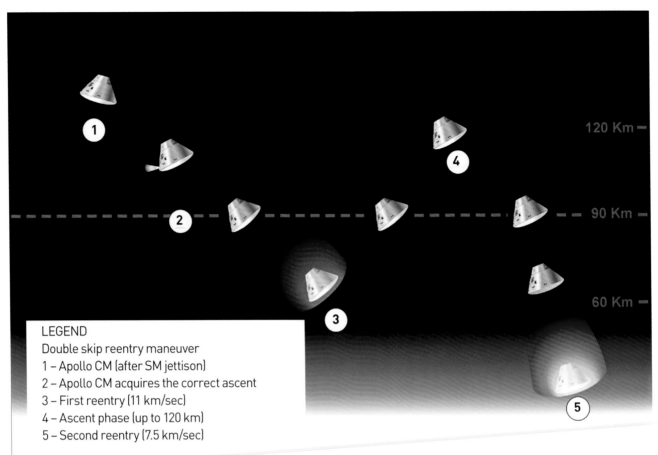

LEGEND
Double skip reentry maneuver
1 – Apollo CM (after SM jettison)
2 – Apollo CM acquires the correct ascent
3 – First reentry (11 km/sec)
4 – Ascent phase (up to 120 km)
5 – Second reentry (7.5 km/sec)

120 Km
90 Km
60 Km

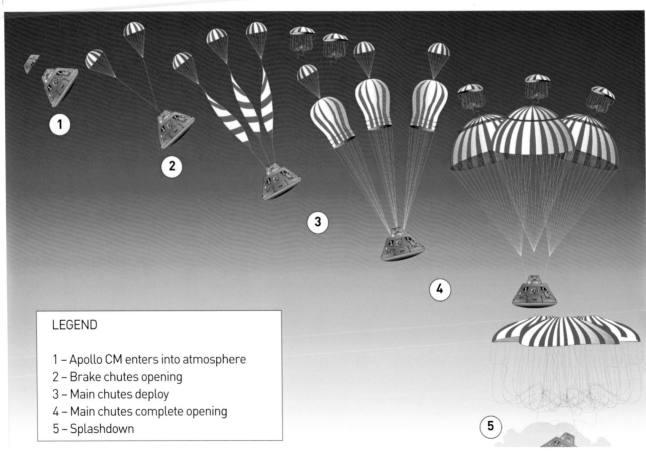

LEGEND

1 – Apollo CM enters into atmosphere
2 – Brake chutes opening
3 – Main chutes deploy
4 – Main chutes complete opening
5 – Splashdown

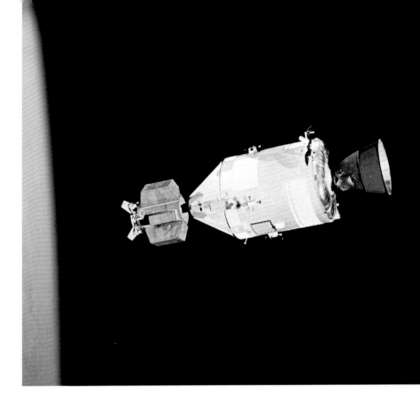

The lunar module multiplied the difficulties of the mission. Fabricated by the Grumman Corporation, it started late, ran behind schedule, and cost more than anticipated. But it also represented the Apollo system's most indispensable component, and at the same time the part most subject to catastrophic failure. Not only did it need to uncouple flawlessly from the command and service module as the two circled the moon, it then depended solely on its single descent engine for a precise, safe landing. On the moon, the lander served as the base of operations for the astronauts. When the time came to go, the lunar module split itself in two, leaving the lower stage on the surface and relying on an ascent stage—powered by just one, nonredundant engine—for the flight back to the orbiting command module. To realize these imposing objectives, the Grumman team found itself beset with control, guidance, and maneuverability problems.

The multiple spacecraft that traveled as one posed the perennial problems of size and weight. The command and service module were shaped like a cylinder with a cone at one end and an engine mount at the other. They measured 36.2 feet (11 meters) long, 12.8 feet (3.9 meters) in diameter, and contained 218 cubic feet (6.2 cubic meters) of interior space. The command module by itself accounted for 10.6 inches (3.2 meters) of the entire length. It weighed 32,390 pounds (14,690 kilograms). In contrast, the lunar module, which looked like a tall, spidery, four-legged creature, stood about 23 feet (7 meters) high and had a width and a depth of 31 feet (9.4 meters). The lunar module weighed 36,200 pounds (16,400 kilograms).

Once in Earth orbit, the spacecraft needed to realign itself for the trip to the moon. The shroud atop the Saturn V that protected the Apollo vehicle during liftoff opened, after which the command and service module detached itself, turned around to face the lunar module still embedded in the shroud, docked with it, and extracted it. As it headed off to the moon, the full spacecraft made for an impressive sight, about 59 feet (nearly 18 meters) from end-to-end and weighing nearly 68,600 pounds (over 31,116 kilograms).

The space agency planned twelve crewed missions for Apollo. The first, Apollo 1, ended before it began. On January 27, 1967, Virgil Grissom, Edward White, and Roger Chaffee died in a fire—the result of an electrical short-circuit igniting the pure oxygen environment in the capsule—on Kennedy Pad 34 as they carried out a launch rehearsal. Under intense criticism from the media and Congress, NASA took twenty-one months before it attempted another manned flight: Apollo 7 in October 1968. The mission successfully tested the command and service module, and the crew practiced withdrawing the lunar module from the Saturn V shroud. Apollo 8 in December 1968 represented

a leap in the program's progression, and an attempt to make up time lost after Apollo 1. Using the Saturn V for the first time with human passengers, it sent Frank Borman, James Lovell, and William Anders on a successful lunar circumnavigation. The Apollo 9 mission in March 1969 flew only in Earth orbit, but practiced the separation, rendezvous, and docking maneuvers of the command and lunar modules. The next flight, Apollo 10 in May 1969, flew astronauts Thomas Stafford, Eugene Cernan, and John Young to orbit the moon and to rehearse the flight plan of the first lunar landing.

The fulfillment of President Kennedy's lunar ambitions occurred in July 1969, with just five months to spare. But Apollo 11 did not go without incident. On the July 20, as Neil Armstrong and Buzz Aldrin flew to the moon's surface in the lunar module (while Michael Collins orbited in the command module), Armstrong took the controls after realizing that the designated touchdown point looked too rock-strewn for a safe landing. He searched the landscape for some time, and with only 2 percent of the descent engine's propellant remaining, he found a spot and made contact. The two men then stepped out, spent two and a half hours collecting almost 48 pounds (21.75 kilograms) of geological samples, and returned to the lunar module for the rendezvous with Collins and the trip home.

The next lunar mission (Apollo 12 in November 1969) unfolded uneventfully, but Apollo 13 in April 1970 almost ended in disaster when, after a routine mechanical stirring of oxygen tank 2 in the service module, an explosion all but incapacitated the spacecraft. The crew retreated to the lunar module, where they waited in cold, hunger, and sleeplessness for the lunar circumnavigation and then the return home. They faced a highly problematic reentry in which only a near-perfect trajectory would prevent the capsule to penetrate the Earth's atmosphere. The crew splashed down safely in the South Pacific on April 17, 1970.

The remainder of the Apollo flights (numbers 14, 15, 16, and 17, in February and July 1971 and in April and December 1972, respectively) happened without major incidents (except for a "wave-off" during the landing sequence in Apollo 16 as John Young and Charles Duke descended to the moon). During the final three voyages, a 10-foot-long (3.1-meter) lunar rover made by Boeing and General Motors gave the astronauts greatly improved mobility, as well as the capacity to find and transport rock and soil samples over greater distances.

In the end, Apollo succeeded as President Kennedy hoped it might. It still represents perhaps the greatest achievement of the Space Age. But it cost dearly too, in the lives of the three Apollo 1 astronauts. It also set an impossible standard of success for future NASA missions.

Vostok 3K

Long before the Space Age began, a handful of idealists dreamed of launching human beings beyond the Earth's atmosphere. The concept gained ground during the early part of the twentieth century when Russian scientist Konstantin Tsiolkovsky (1857–1935), German/Romanian physicist Hermann Oberth (1894–1989), and American physicist Robert H. Goddard (1882–1945) propagated rocketry and human spaceflight.

Sergei P. Korolev (1906–1966) did more than any single individual to realize this vision. Korolev served as the chief designer for the USSR's Special Design Bureau 1 (OKB-1) and became perhaps the world's leading proponent and practitioner of modern spaceflight. As early as the late 1940s and early 1950s, Korolev's team developed a plan to send a piloted rocket on a suborbital ballistic arc up to 124 miles (200 kilometers) long.

Tikhonravov's spacecraft, known as Vostok ("east" in Russian), consisted of two parts: a descent module/pilot cabin and an

Vostok/Voshod Family Development

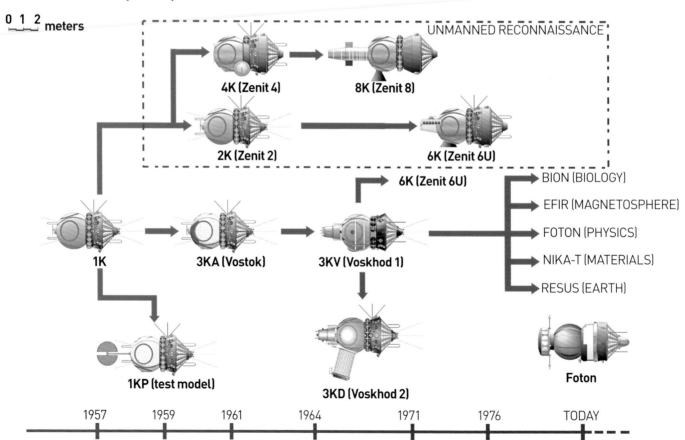

0 1 2 meters

UNMANNED RECONNAISSANCE

4K (Zenit 4)   8K (Zenit 8)

2K (Zenit 2)   6K (Zenit 6U)

6K (Zenit 6U)

BION (BIOLOGY)

EFIR (MAGNETOSPHERE)

FOTON (PHYSICS)

NIKA-T (MATERIALS)

RESUS (EARTH)

1K   3KA (Vostok)   3KV (Voskhod 1)

1KP (test model)

3KD (Voskhod 2)

Foton

1957   1959   1961   1964   1971   1976   TODAY

Vostok 3KA

0   1   2
|—·—|——|——·—| meters

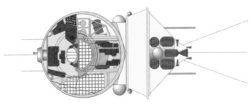

TOP SECTION

FRONT SECTION

SIDE SECTION

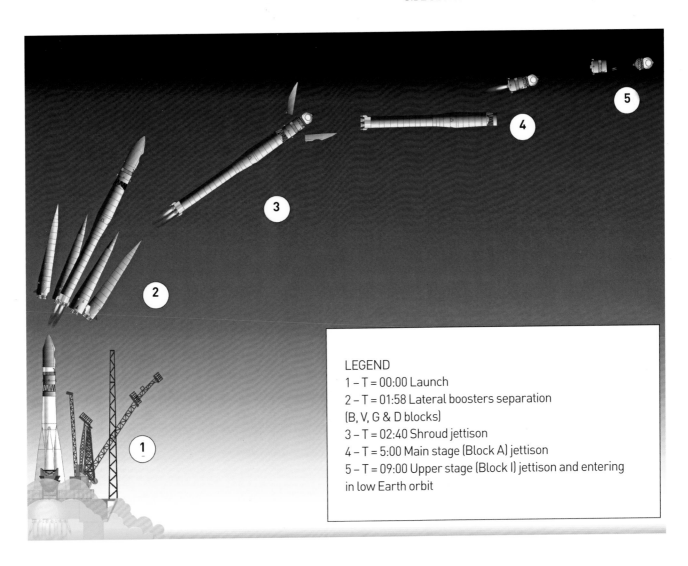

LEGEND
1 – T = 00:00 Launch
2 – T = 01:58 Lateral boosters separation
(B, V, G & D blocks)
3 – T = 02:40 Shroud jettison
4 – T = 5:00 Main stage (Block A) jettison
5 – T = 09:00 Upper stage (Block I) jettison and entering
in low Earth orbit

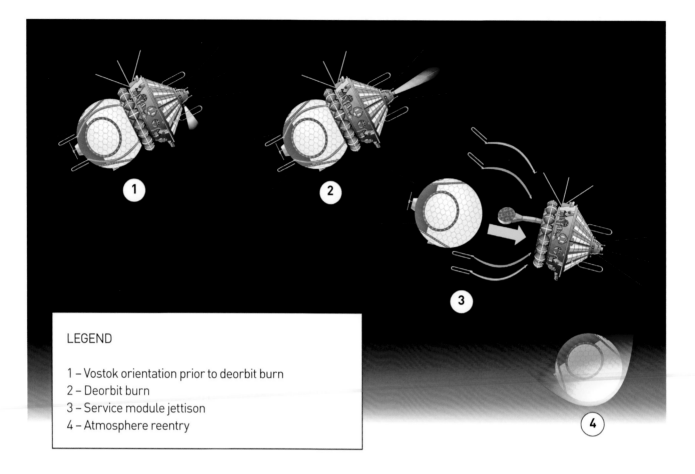

instrument module with a braking system. Korolev's bureau served as the prime contractor for Vostok, although fifteen other organizations made significant contributions to the project. He also oversaw the modification of the R-7 ballistic missile as the launch vehicle for the Vostok missions, increasing its power sufficiently to boost the heavy spacecraft into near-Earth orbit. This rocket became known as the two-stage Vostok-K.

Among the Vostok series, the Vostok 3KA occupies a special prominence among the world's space capsules, even though it looked deceptively simple. Like a ball and a cone stuck together incongruously, the spherical portion constituted Vostok's reentry module. It measured about 7.5 feet (2.3 meters) in diameter and weighed 5,423 pounds (2,460 kilograms). The cone, or equipment module, had a length of about 7.4 feet (2.25 meters), a diameter of nearly 8 feet (2.43 meters), and a mass of 5,004 pounds (2,270 kilograms).

Vostok 3KA gained its reputation by being the first spacecraft to transport a human being into space. Before that milestone, Vostok flew several unmanned missions: in May 1960, a flight that failed during recovery; in August of the same year, a mission that

carried two dogs (Belka and Strelka) and returned successfully after eighteen orbits; in December 1960, an attempt that ended due to another recovery system failure; and in March 1961, two launches on which canine travelers survived the rigors of space.

After these trials, the Soviet space agency planned a voyage that, if successful, promised to catapult the USSR further ahead in the space race. (To this point, NASA had flown no crewed space flights of any kind and would have none until Alan Shepard flew a brief suborbital mission on May 5, 1961). The Soviets selected Yuri A. Gagarin, an outgoing and amiable air force first lieutenant who would make a good impression on the world's media. Vostok 1 lifted off from Baikonur on April 12, 1961. After a single orbit and eighty-nine minutes in space, Gagarin landed in the region of Rostov, near the village of Smelovka. The Soviet government and people celebrated him with a hero's return and he embarked on tours of many foreign capitals.

If the news of Gagarin did not cause despair in the American space agency and in the US political establishment, what followed in the succeeding Vostoks certainly aroused great concern. Vostok 2 went into orbit in June 1961, during which flight

Gherman Titov made seventeen revolutions of the Earth. During the next year, in August, Vostoks 3 and 4 lifted off with Andrian Nikolayev making sixty-four orbits and Pavel Popovich, forty-eight, respectively. Finally, Vostoks 5 and 6 went into space in June 1963, with Valeri Bykovsky achieving a record eighty-one Earth revolutions (almost five days in space) and the first woman cosmonaut, Valentina Tereshkova, completing forty-eight. The month before, on May 15 and 16, 1963, astronaut L. Gordon Cooper flew Mercury's last mission (Mercury-Atlas-9, or Faith-7) for twenty-two orbits, spending less than a day and a half aloft.

Voskhod 3KV

Soviet superiority in space—so evident in the dual between Vostok and Mercury—started to falter in the subsequent contest between Voskhod and Gemini. The year 1966 marked a pivot point: the completion both of Gemini and Voskhod, and more decisively, the year in which Sergei Korolev, the indispensable man in the Soviet program, died accidentally after surgery at age fifty-nine. Voskhod had at least two points of origin. On the technical side, Korolev wrote as early as January 1963 about the technical value of long-duration flights that carried scientists trained as cosmonauts. To facilitate such missions, he pressed for multimember crews.

Voskhod 3KV (Voskhod 1)

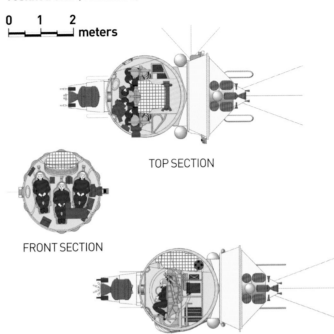

0   1   2
meters

TOP SECTION

FRONT SECTION

SIDE SECTION

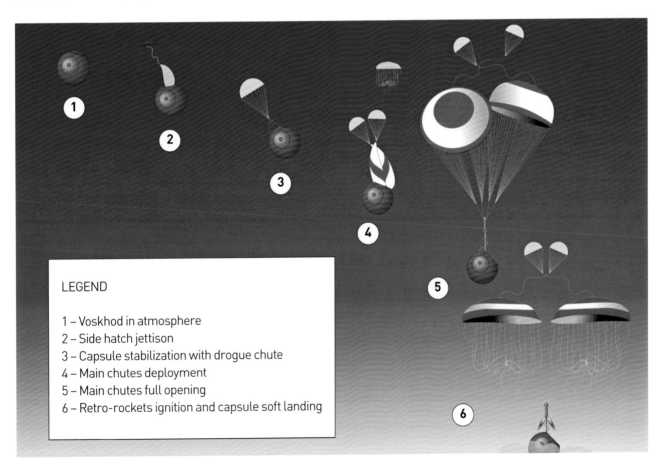

LEGEND

1 – Voskhod in atmosphere
2 – Side hatch jettison
3 – Capsule stabilization with drogue chute
4 – Main chutes deployment
5 – Main chutes full opening
6 – Retro-rockets ignition and capsule soft landing

Politics dominated the second reason for increasing the number of crewmembers. By late 1963, American plans for two-seat Gemini and three-seat Apollo spacecraft became known, pressuring Korolev and the Soviet space program to keep pace. A new Soyuz spacecraft in the development phase could not be finished fast enough to meet the US challenge, so in February 1964 Korolev took four Vostok already in the pipeline and directed his team to refashion their interiors as three-seaters, giving them the name Voskhod, or Sunrise. Preliminary specifications rolled out in April 1964, the same month that the project got the official go-ahead. Under the duress of time, the Voskhod design group found itself improvising and making rash decisions. To pack in three men and save space, the engineers eliminated ejection seats, and with them any means of emergency crew escape at launch or landing. They instead planned for a hard touchdown with the passengers remaining inside of the capsule, the impact of which would be broken by a small cluster of solid rockets attached to a second main parachute. By removing the ejection seats and taking other measures, three cosmonauts could be accommodated; but the instrument panel stayed in its original place, requiring the Voskhod crew to crane their necks to see the dashboard. And Korolev and his designers further cut down bulk by eliminating space suits, relying only on cabin atmosphere to sustain the cosmonauts.

The Voskhod 3KV emerged from this welter of compromises. Like its brother ship, the Vostok, the 3KV had the ungainly look of a sphere mated to a cone. It measured 16.4 feet (5 meters) long, nearly 8 feet (2.4 meters) in maximum diameter, and weighed a total of 12,527 pounds (5,682 kilograms). The Voskhod 11A57 launch vehicle (another R-7 variant) consisted of a Vostok-K with a more powerful upper stage borrowed from the Molniya booster, enabling it to lift the extra 2,000 pounds gained from transforming the Vostok into the Voskhod.

The initial flights of the Voskhods occurred in October 1964. First came a trial run, designated Kosmos 47, on the sixth of the month, which lasted less than a day and tested the spacecraft's systems. Then, on the twelfth, pilot Vladimir Komarov, scientist Konstantin Feoktistov, and physician Boris Yegorov embarked on Voskhod 1, the world's first spaceflight with more than one passenger on board. They made sixteen orbits in a little more than twenty-four hours. The flight devoted itself to biomedical research and to the practicalities of integrating the activities of a multidisciplinary crew.

The last Voskhod 3KV mission flew in March 1966 under the name Kosmos 110. Two dogs—Veterok and Ugolyok—orbited for twenty-two days in a biological experiment that sent them through the Van Allen radiation belts. Both landed safely.

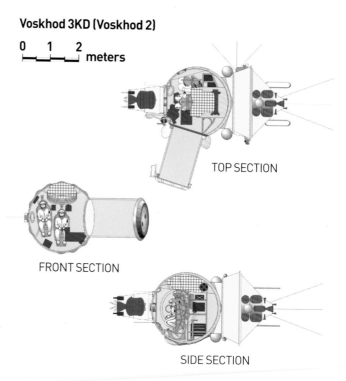

Voskhod 3KD (Voskhod 2)

0   1   2 meters

TOP SECTION

FRONT SECTION

SIDE SECTION

Voskhod 3KD

Not content just to be the first to fly three human beings into space on the Voskhod 1 mission, the team that built the Voskhod 3KV—led by Sergei Korolev, the chief designer of Special Design Bureau 1—attempted an even more daring adventure in the mission of Voskhod 2. Little separated the 3KV from the new capsule, known as the 3KD. They had similar dimensions and weighed about the same. They used the same Voskhod 11A57 booster. To conserve space and add passengers, neither came equipped with crew escape systems or ejection seats. Only two cosmonauts flew on Voskhod 2, so the third seat could be removed. And, unlike Voskhod 1, the cosmonauts wore space suits. The 3KD also carried the Volga airlock, a 551-pound (250-kilogram) inflatable add-on not available on the 3KV.

A Voskhod 2 trial flight happened in February 1965, when an unmanned Voskhod 3KD tested the Volga airlock successfully. Then, on March 18, 1965, the Voshkod 2 mission took off from the Baikonur Cosmodrome with two cosmonauts and a high-risk objective: to conduct the first spacewalk ever attempted. As soon as pilot Alexey Leonov and Commander Pavel Belyayev reached orbit, they attached a backpack containing oxygen to Leonov's suit and Belyayev pressurized the inflatable airlock. Leonov then stepped into it, Belyayev sealed it behind him and depressurized it, and on the second orbit, over the Black Sea, Leonov opened the Volga's hatch and moved away from the capsule 17.6 feet (5.35

Voskhod 3KD (Voskhod 2)

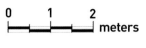

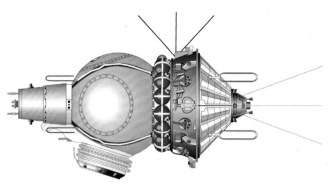

TOP VIEW

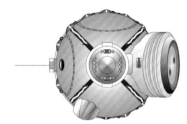

FRONT VIEW

SIDE VIEW

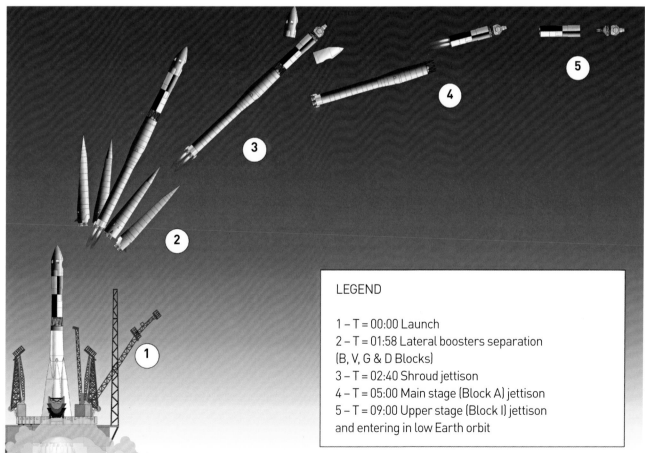

LEGEND

1 – T = 00:00 Launch
2 – T = 01:58 Lateral boosters separation
(B, V, G & D Blocks)
3 – T = 02:40 Shroud jettison
4 – T = 05:00 Main stage (Block A) jettison
5 – T = 09:00 Upper stage (Block I) jettison
and entering in low Earth orbit

Soyuz 7K–A (Early Soyuz)

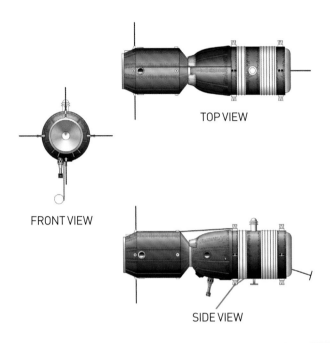

TOP VIEW

FRONT VIEW

SIDE VIEW

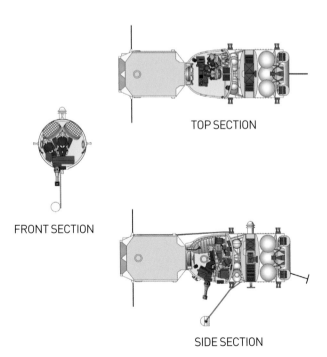

TOP SECTION

FRONT SECTION

SIDE SECTION

0 1 2 3 4 5 meters

meters): the umbilical's maximum length. Leonov accomplished two scheduled tasks—he attached a camera to the airlock to record his actions and took photographs of the spacecraft. He remained outside for a little more than twelve minutes. At one point, Leonov pulled the tether too sharply and had to brace himself to avoid colliding with the Voskhod.

Then serious problems occurred. When Leonov attempted to return to Voshkod through the airlock, he found that his suit had swelled to such an extent that its stiffened joints prevented him from bending his body adequately to reenter. Realizing the gravity of the situation, he struggled with all his might for eight minutes, even taking the dangerous step of bleeding off air from the suit to shrink its bulk, which dropped the pressure below safe limits. Leonov finally forced his way back in, only to find that he and Belyayev could not get the hatch to seal properly, the consequence of thermal distortion brought on by Leonov's battle at the airlock. They eventually succeeded in closing the hatch.

But these harrowing events did not end there. On the sixteenth and last orbit, the spacecraft's automatic reentry system failed. The cosmonauts flew an extra orbit so that Belyayev could familiarize himself with the procedures for a manual landing. But during the forty-six seconds that elapsed from the time Leonov and Belyayev reoriented the spacecraft for reentry to the time it took for them to stumble back to their seats in bulky spacesuits, Voskhod's center of mass shifted, causing the landing point to go off target. When Leonov and Belyayev touched down, the Voskhod sank into heavy snow, about 1,243 miles (2,000 kilometers) from the expected landing site, in a forest near Perm. They waited for more than two hours for a rescue helicopter to collect them.

The hair-raising journey marked the peak of Soviet preeminence in space up to that point.

Soyuz 7K-OK

When Sergei Korolev led his engineers in developing Voshkod, he and his team also designed the first Soyuz ("union") spacecraft. The Soyuz series coincided with the earliest attempts by the Soviets to respond to the US Apollo program. Specifically, Korolev's engineers mated the Vostok spacecraft with a crew-carrying moon capsule known as the IL, designed for lunar circumnavigation. The Vostok-7/IL concept failed to win practical support, but it survived inside of OKB-1, to be amalgamated with an orbital spacecraft, Sever. Completed by March 1963, the combined vehicle came to be designated as Soyuz 7-KA, the first of a family of spacecraft that continues to fly more than fifty years later.

Korolev's team began work on the 7-KA's successor, the 7K-OK, in spring 1963 and won approval for it in December

Soyuz 7K-OKS (Active Docking)

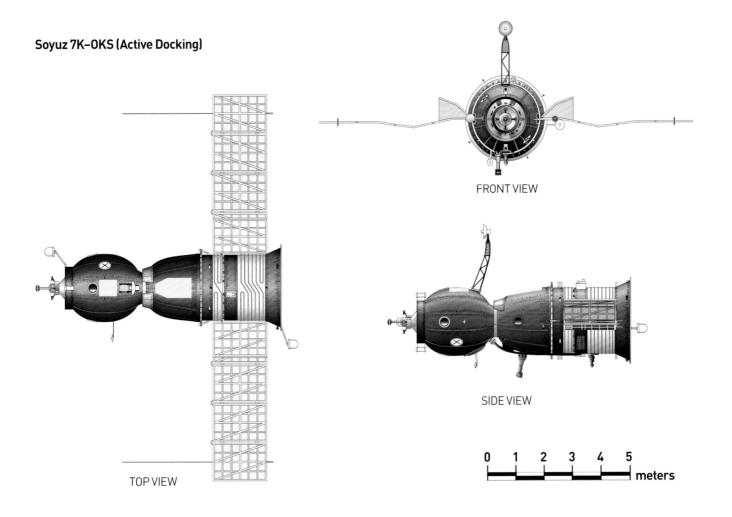

FRONT VIEW

SIDE VIEW

TOP VIEW

0 1 2 3 4 5 meters

of that year. The team envisioned a vehicle that offered such innovations as maneuvering in orbit and automated rendezvous, approach, and docking. Korolev originally intended the 7K-OK to fulfill the first leg of a lunar mission, to be used in conjunction with another of his designs, the 7K-L1. In this vision, the 7K-L1 would be launched into low Earth orbit, and rendezvous with the crew-carrying 7K-OK, at which point the cosmonauts would transfer to the 7K-L1 for a circumnavigation of the moon. But under pressure from NASA's lunar program, the Soviets abandoned Korolev's concept in favor of a lunar orbit rendezvous architecture similar to Apollo's: the 7K-LOK capsule, the LK lunar lander, and the N1 rocket.

Because the Soviet program ultimately lost to the US in the race to the moon, the 7K-OK found a less lofty but pragmatic purpose nonetheless—as a conveyance for astronaut crews to orbit, where they docked with the first generation of space stations. Practically, however, the 7K-OK proved to be a troubled

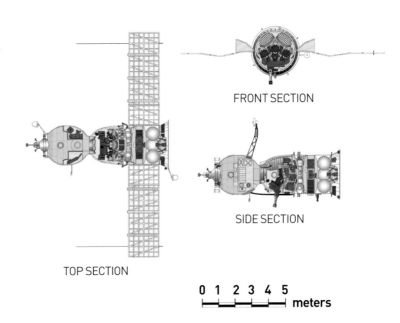

FRONT SECTION

SIDE SECTION

TOP SECTION

0 1 2 3 4 5 meters

Soyuz 7K–OK (Active Docking)

0 1 2 3 4 5 meters

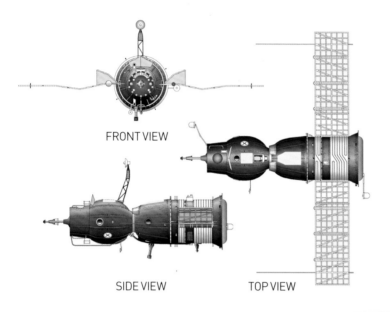

FRONT VIEW

SIDE VIEW

TOP VIEW

program. The first unmanned launch occurred in November 1966 in a mission that uncovered extensive construction deficiencies, resulting in an uncontrollable spacecraft that had to be destroyed during its flight. The following month, another failure: the launch escape system activated while the 7K-OK and its launch vehicle sat on the pad, pulling the capsule away, but causing the rocket to explode, killing and injuring several people nearby. A third flight sent the 7K-OK into the Aral Sea. Worst of all, in April 1967, cosmonaut and test pilot Vladimir Komarov died during the landing of the fourth mission. After a redesign, the 7K-OK went into orbit with and without crews thirteen more times, with relative success. The program ended in 1971.

Despite its checkered record, the 7K-OK originated the famous tripartite design common to the Soyuz family. At the forward end, a spherical habitation module offered the cosmonauts essential life support functions, including toilets, that enabled them to stay aloft for much longer periods than in the Voskhod. It also served as an airlock. The midsection consisted of a bell-shaped descent module for the reentry part of the mission.

Soyuz 7K–LOK

0 1 2 3 4 5 meters

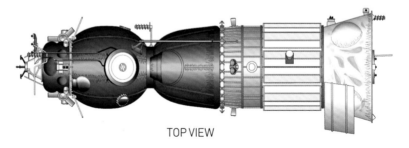

TOP VIEW

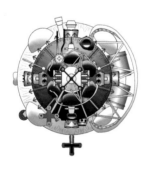

FRONT VIEW

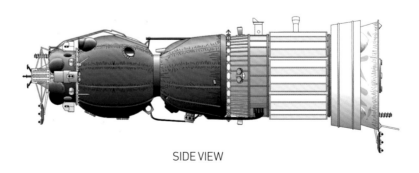

SIDE VIEW

Finally, aft of the descent module, a cylindrical portion known as the service module carried the spacecraft's communications, avionics, guidance, and telemetry, in addition to housing the approach and attitude control thrusters, the antennae, and the solar arrays. The 7K-OK measured just under 25 feet (7.6 meters) long with a maximum diameter of almost 7.5 feet (2.72 meters) and a solar array span of about 27.5 feet (8.37 meters). Once launched, the 7K-OK weighed between 14,220 and 14,462 pounds (6,450 to 6,560 kilograms), depending on the mission.

Following its uneven career, the 7K-OK underwent modifications, eventually emerging as the 7K-T, a safer and more reliable space ferry that brought cosmonauts to the Almaz and Salyut space stations.

Soyuz 7K-OKS

After several unsuccessful attempts to close the gap with the US in the race to the moon, the Soviet space agency abandoned the effort in the early 1970s and reoriented itself toward a new objective: long-duration spaceflight. This redirection became evident in April 1971, when the Soviets placed in orbit the world's first space station, Salyut 1. To transport crews and supplies to Salyut, Special Design Bureau 1 modified its Soyuz 7K-OK spacecraft for this purpose.

Launched just two times, the 7K-OKS managed to be both effective and catastrophic at the same time. It flew for the first time on April 23, 1971, on the Soyuz 10 mission, three days after Salyut 1 went into orbit. During the approach to the station, Commander Vladimir Shatalov made a manual docking attempt. The two spacecraft connected, although he could not get the correct angle of approach for a hard lock. The OKS stayed in place for five and a half hours, but the crew did not enter the station due to a faulty hatch on the OKS. With some difficulty, the Soyuz detached itself for the return to Earth and made a safe landing despite toxic fumes that fouled the air supply.

On the Soyuz 11 mission, the OKS seemed to prove itself—until the end. Commander Georgy Dobrolovsky, Flight Engineer Vladislav Volkov, and Test Engineer Viktor Patsayev went into orbit on June 6, 1971. This time the manual docking occurred as planned. They spent twenty-three days on Salyut 1 preparing it for habitation, and at the end, they transferred their research materials and logs, undocked without incident, and departed for home. Although the crew lost all communications with mission control once the braking engine fired, the automated reentry proceeded smoothly, with an on-target touchdown. But when ground technicians opened the hatch, they found all three men dead. An investigation discovered that a faulty valve had caused a fatal decompression in the cabin. This ended the short career of the Soyuz 7K-OKS.

Soyuz 7K–L1 (Zond 4-8)

0 1 2 3 4 5 meters

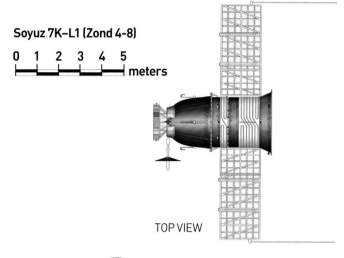

TOP VIEW

FRONT VIEW

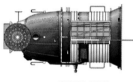

SIDE VIEW

Soyuz 7K-L1 (if manned)

0 1 2 3 4 5 meters

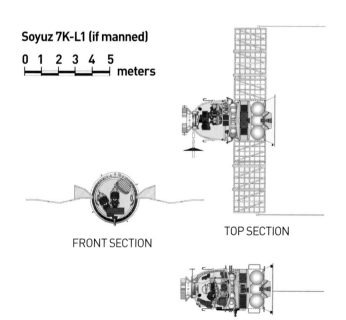

FRONT SECTION

TOP SECTION

SIDE SECTION

Soyuz 7K–LOK

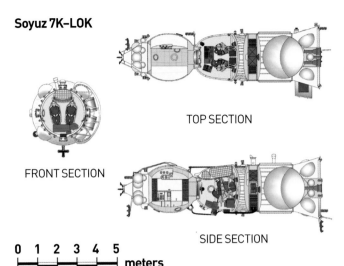

TOP SECTION

FRONT SECTION

SIDE SECTION

0 1 2 3 4 5 meters

Soyuz 7K-L1

Despite his preeminence among Soviet aerospace designers, Sergei Korolev encountered some stiff competition. One such rivalry emerged when, in his role as chief of Special Design Bureau 1 (OKB-1), he and his team proposed a Soyuz-based spacecraft capable of orbiting two cosmonauts around the moon, a belated attempt to catch up to the US lunar program.

For the circumnavigation, Korolev selected the Soyuz 7K-OK to carry the cosmonauts into Earth orbit. Once there, they would rendezvous with his new creation, the 7K-L1, transfer to it, and depart for the moon.

In its circumnavigation role, the 7K-L1 differed from the other Soyuz spacecraft in one main respect: rather than relying on the classic Soyuz three-part structure, Korolev eliminated the habitation module to save weight. He did so because he lacked a rocket powerful enough to carry a full-sized Soyuz to the moon and back. The resulting capsule, although not light or small, weighed roughly 12,500 pounds (5,680 kilograms) and measured 15.7 feet (4.796 meters) long and almost 9 feet (2.72 meters) in diameter.

But one of Korolev's main antagonists—Vladimir Chelomei, the chief designer of OKB-52—had his own ambitions. Chelomei pressed Nikita Khrushchev to reject Korolev's 7K-L1 in favor of his competing LK-1 capsule, which he proposed to send directly to the moon on the large UR-500 Proton rocket. Chelomei's plan won Khrushchev's approval, but with the Soviet premier's ouster in October 1964, Korolev revived the 7K-L1, which won official acceptance in October 1965. No doubt galling to Chelomei, the Soviet space authorities selected his own UR-500 as the launch vehicle for the 7K-L1 since at the time, it

alone generated enough thrust to propel this stripped-down spacecraft directly to the moon.

Korolev and his successor at Special Design Bureau 1, Vasily Mishin, made two additions necessary for the 7K-L1 to be successful. With the habitation module eliminated to reduce the mass of the spacecraft, the 7K-L1 had no mode of egress, so the engineers eliminated the reserve parachute on the reentry module and replaced it with an exit hatch. Second, Korolev's team transferred the N1 rocket's Block D translunar injection stage onto the Proton.

Perhaps the intense pressure to match the Apollo program, and almost certainly the death of Korolev in 1966, caused the Soviet space program to become almost as accident-prone during its lunar project as it had been triumphant during its initial phase. In twelve missions involving the 7K-L1 from March 1967 to October 1970, it experienced eight failures. After testing the Block D in the first flight, six successive attempts went wrong: the Block D in mission 2, the Proton first stage in mission 3, the second stage in mission 4, destruction on reentry in mission 5, escape system initiation on launch in mission 6, and an on-pad explosion of the upper stage in mission 7. Flights 9 and 10 miscarried, respectively, due to a crash on reentry and second stage failure. Only in missions 8, 11, and 12 (September 1968, August 1969, and October 1970, respectively) did the un-crewed 7K-L1 orbit the moon successfully. Faced with many mishaps and the success of Apollo, the Soviet space agency terminated the 7K-L1 in 1970 without mounting any manned flights.

Soyuz 7K-LOK

Although distinctly different from the parallel American spacecraft—the Saturn V, the lunar module, and the command and service module—in technical make-up, the Soviet Soyuz spacecraft borrowed the overall mission architecture (lunar orbit rendezvous, or LOR) from Apollo to save time and have a chance to close the space race gap. In doing so, the Soviets abandoned an earlier plan for Earth-orbit rendezvous, changing instead to a crew capsule and a lunar lander combined in one spacecraft.

And the Soviets also had to do it without their leader, Sergei P. Korolev, who died suddenly in 1966. Before his death, Korolev led his team at OKB-1 in designing the Soyuz 7K-LOK (LOK stood for "lunniy orbitalny korabl," or lunar orbital craft). Korolev submitted an outline for a lunar lander to the Communist Party Central Committee as early as 1960, but not until August 1964 did the authorities make a definitive decision to launch a program to reach the moon—more than three years after President Kennedy's request to Congress for a lunar program.

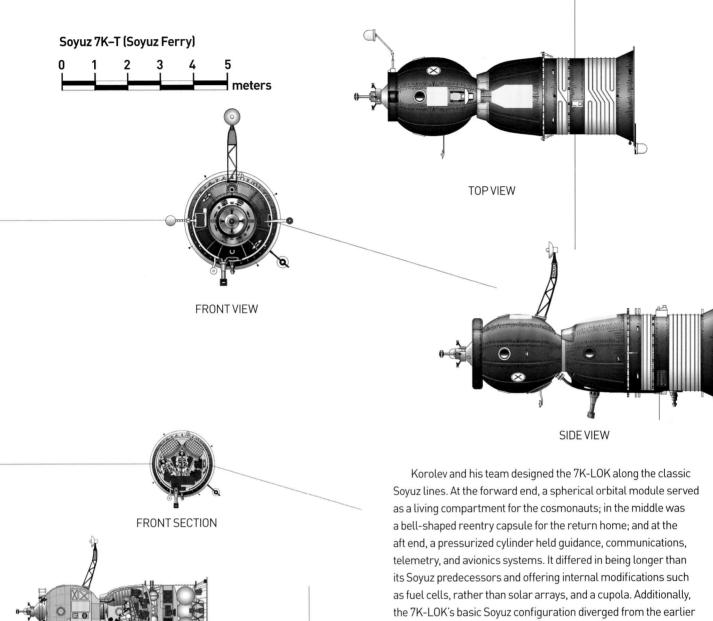

Soyuz 7K–T (Soyuz Ferry)

0   1   2   3   4   5
meters

TOP VIEW

FRONT VIEW

SIDE VIEW

FRONT SECTION

SIDE SECTION

TOP SECTION

0   1   2   3   4   5
meters

Korolev and his team designed the 7K-LOK along the classic Soyuz lines. At the forward end, a spherical orbital module served as a living compartment for the cosmonauts; in the middle was a bell-shaped reentry capsule for the return home; and at the aft end, a pressurized cylinder held guidance, communications, telemetry, and avionics systems. It differed in being longer than its Soyuz predecessors and offering internal modifications such as fuel cells, rather than solar arrays, and a cupola. Additionally, the 7K-LOK's basic Soyuz configuration diverged from the earlier Vostok and Voskhod in that it enabled its crew to conduct active maneuvering, as well as orbital rendezvous and docking. Still, while the 7K-LOK could berth with other spacecraft, it lacked a transfer tunnel, which meant that crews could only exit the 7K-LOK or enter other vehicles by means of spacewalks.

The completed 7K-LOK had imposing dimensions: 33 feet (10 meters) long and 9.6 feet (2.9 meters) in diameter. It weighed 21,720 pounds (9,850 kilograms) in lunar orbit.

The Korolev design bureau manufactured six 7K-LOKs. The first, a dummy, went into low Earth orbit in December 1970 aboard a Proton rocket. During the next two missions in June 1971 and November 1972—another dummy test, followed by a simulated operational flight, both intended as lunar flybys—

Soyuz 7K–M (Soyuz ASTP)

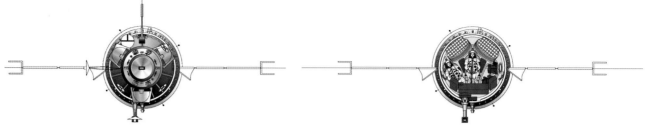

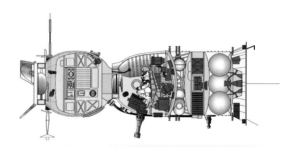

FRONT VIEW

FRONT SECTION

SIDE VIEW

SIDE SECTION

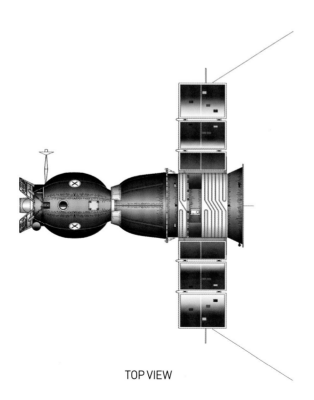

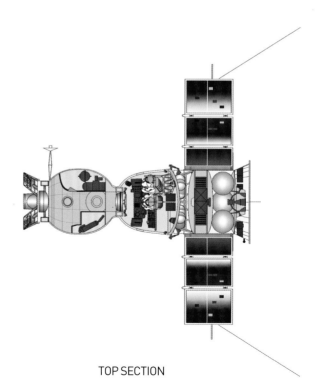

TOP VIEW

TOP SECTION

0 1 2 3 4 5 meters

the N1 boosters failed. Although ambitious mission planners scheduled full circumlunar expeditions for August 1974, with manned flights to follow, the Soviet space agency cancelled the N1 and 7K-LOK programs in May 1974.

The Soviets ultimately made a wise decision about their loss to Apollo. Rather than persist, they changed course, and with the launch of Salyut 1 in April 1971, they initiated the world's first space station, a specialty they pursued with much success for decades. And in support of this initiative, the Soyuz spacecraft assumed an indispensable role as crew taxi to and from the Salyuts and their successors.

Soyuz 7K-T

After the deaths of three cosmonauts in April 1971 aboard the Soyuz 7K-OKS, during mission 11 to the Salyut 1 space station, the Soviet space agency knew it needed to make improvements but stuck to the basic 7K-OKS concept. The Soviet devotion to incremental change did not waver, even in the face of such a heavy blow. The modifications resulted in the Soyuz 7K-T. Because the crewmembers of the 7K-OKS died from cabin depressurization, engineers at Special Design Bureau 1 decided to reduce the crew complement from three to two in the Soyuz 7K-T to allow room for the cosmonauts to wear pressure suits during the key danger points in the missions: launching, docking, undocking, and reentry. The space saved by this measure also accommodated extra life-support systems. Another improvement involved the replacement of the solar panels on the OKS with batteries. Although the batteries lasted only two days, they could be recharged once the 7K-T docked with the Salyut. Additionally, for missions to the three Almaz military space stations (designated Salyuts 2, 3, and 5 to disguise their true identities), the Soyuz 7K-T underwent further changes that entailed remote control of the station and the installation of a new parachute system. To distinguish it from the standard 7K-T, the Soyuz that serviced Almaz became known as the 7K-T/A9.

Two unmanned test flights of the 7K-T preceded its operational career. The first went into space in June 1972, and it stayed aloft for a week. Then, the Soyuz rocket used for the 7K-T launches failed to put a Zenit reconnaissance satellite into orbit, which delayed the next 7K-T liftoff for a year pending modifications of the booster. Finally, the second test 7K-T flew in June 1973 and spent two days in orbit.

With the launch of the 7K-T on September 27, 1973, during Soyuz mission 12, the USSR attempted its first manned spaceflight since the 7K-OKS accident. Essentially a shakedown cruise, the spacecraft underwent systems tests by cosmonauts Vasili Lazarev and Oleg Makarov. They landed after two days and

reported that it performed well. With that, the 7K-T began a long and largely successful life providing crew transport to succeeding Salyut space stations.

Soyuz 7K-M (ASTP)

Soyuz 7K-M embodied the latest incarnation in a series of spacecraft dating back to 1971. As part of the Soviet's pursuit of space stations and long-duration spaceflight, the Soyuz 7K-M was a modified version of the Soyuz 7K-T, the 7K-M modified for docking with the Apollo capsule during the **Apollo–Soyuz Test Project** (ASTP). The ASTP was the first joint US–Soviet spaceflight.

For its part, the Soyuz 7K-M weighed 14,720 pounds (6,680 kilograms) fueled and measured 24.5 feet (7.5 meters) long, with a diameter of 8.9 inches (2.72 meters), and a span of 27.5 feet (8.37 meters).

The modifications that recast the Soyuz 7K-T into the 7K-M all concerned compatibility issues with the American spacecraft. The Soyuz got new, lighter solar panels for longer mission life, radio aerials for common communications, optical targets for docking manually with Apollo, and changes in the environmental controls to lower the cabin pressure. But the major retrofit involved a new docking mechanism, which replaced the existing Soyuz male configuration with an androgynous universal system, effective in active or passive modes. Conceived and fabricated in the United States, it underwent testing by NASA and Soviet engineers in an early act of international cooperation between the space race rivals.

This unique apparatus proved to be one of the most noteworthy technical advances made during ASTP, one that NASA appropriated for its own uses when it imported the androgynous docking system from the Apollo-Soyuz Test Program into the design of the space shuttle.

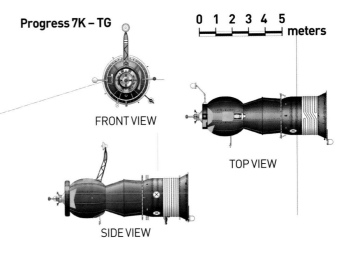

Progress 7K – TG

0 1 2 3 4 5 meters

FRONT VIEW

TOP VIEW

SIDE VIEW

Progress 7K-TG

The Soviet space agency's focus on long-duration spaceflight rested on the development of a reliable vehicle to haul supplies to its space stations. Without regular replenishment, no station could be successful. Special Design Bureau 1 considered converting the basic Soyuz 7K-OK spacecraft into an automated space truck for the Salyut stations. Rather than pursue that option, it instead made an aborted attempt in the late 1960s to build a military version of the Soyuz known as the 7K-G. By mid-1973, the Korolev bureau began work on a true supply ship known as the 7K-TG (a designation based on the Russian language equivalent of "transport cargo"). Designers patterned it on the Soyuz 7K-T, a successful crew-ferrying spacecraft.

In fabricating the 7K-TG, project engineers concentrated their modifications on the Soyuz 7K-T's bell-shaped descent module. They removed the accoutrement required for human passengers, installed the tanks and pumps necessary to refuel the station automatically once it docked, and created storage space for supplies, food, and fuel. They also eliminated the solar arrays in favor of chemical batteries. And in addition to fuel and supplies, they enabled 7K-TG to carry such mission-oriented apparatuses as a KRT-10 radio telescope for Salyut 6 and the Iskra 3 satellite for Salyut 7 (later deployed from the station). Once empty, the 7K-TG served as a refuse bin that disposed of the station's waste by detaching itself and burning up on reentry into the atmosphere.

Although built to operate in free flight for up to three days and to dock for as long as thirty-three, in practice the later Progress models orbited with the stations for as long as seventy-five days. Perhaps in a symbolic recognition that the 7K-TG was a practical, reliable, and indispensable ingredient in the Soviet goal of a permanent presence in space, authorities eventually renamed it Progress 7K-TG, the first in a series that included Progress-M and later Progress-M1.

Progress 7K-TG measured 24.5 feet (7.48 meters) in length, nearly 12 feet (2.72 meters) in diameter, and weighed 15,480 pounds (7,020 kilograms) loaded with fuel. It delivered payloads of about 5,100 pounds (2,313 kilograms), depending on requirements. Progress 7K-TG served for more than twelve years (January 1978 to May 1990), dividing its missions among three space stations: going twelve times to Salyut 6, thirteen times to Salyut 7, and eighteen times to Mir.

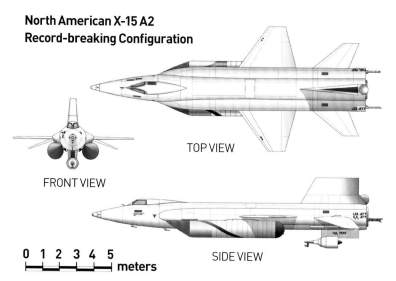

North American X-15 A2
Record-breaking Configuration

TOP VIEW

FRONT VIEW

0 1 2 3 4 5 meters

SIDE VIEW

SPACEPLANES
X-15

Despite the common belief that America's space program began solely as a byproduct of the Cold War rivalry between the United States and the USSR, it actually started several years earlier, the result not of political forces, but of the onrush of technology.

America opened a path to space travel on October 14, 1947, when Capt. Chuck Yeager flew the Bell X-1 aircraft faster than the speed of sound. This feat represented a collaboration between the US Air Force and the National Advisory Committee for Aeronautics (NACA), a federal agency created in 1915 to advance aeronautical research. After the Mach 1 milestone, the pace of high-speed flight accelerated. Just six years later, on December 3, 1953—fifty years to the day since the Wright brothers achieved the first powered, heavier-than-air flight—NACA's Scott Crossfield flew the Douglas D-558 Skyrocket aircraft over Mach 2.

Success in these two projects led the NACA's engineers to believe that no insurmountable technological barriers stood in the way of *hypersonic* flight (Mach 5 and above). The director of the NACA, Dr. Hugh L. Dryden, enlisted the help of high ranking navy and air force research and development officials to reach this goal. In 1954, he persuaded them to fund the fabrication of three hypersonic prototype aircraft designed by the NACA, and to test them under NACA supervision. The three agencies signed a memorandum of understanding in December 1954, initiating Air Force Project 1226—better known as the X-15 aircraft.

The X-15 concept originated with the hypersonic aircraft committee at the NACA's Langley Memorial Aeronautical Laboratory in Hampton, Virginia. Langley's design envisioned

North American X-15 A2
Record-breaking Configuration

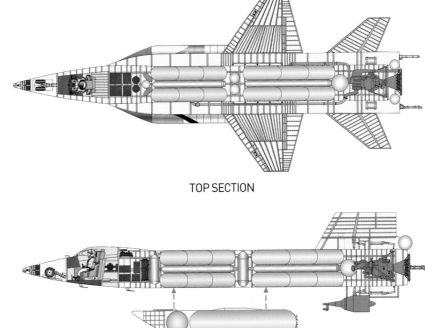

TOP SECTION

FRONT SECTION

SIDE SECTION

0 1 2 3 4 5 meters

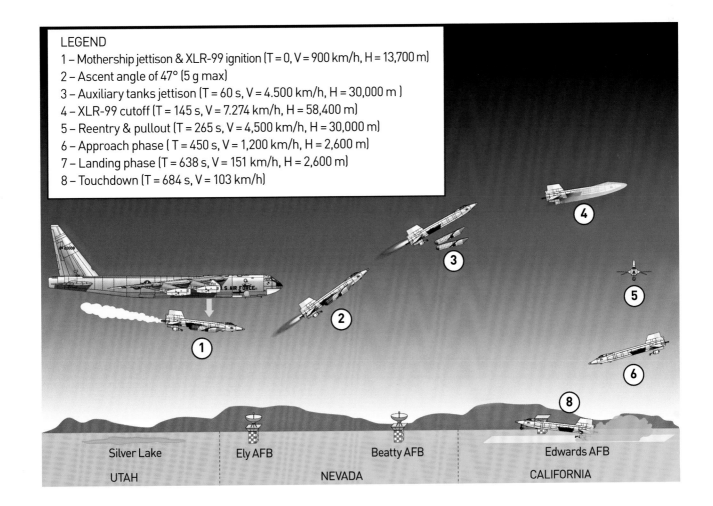

LEGEND
1 – Mothership jettison & XLR-99 ignition (T = 0, V = 900 km/h, H = 13,700 m)
2 – Ascent angle of 47° (5 g max)
3 – Auxiliary tanks jettison (T = 60 s, V = 4.500 km/h, H = 30,000 m)
4 – XLR-99 cutoff (T = 145 s, V = 7.274 km/h, H = 58,400 m)
5 – Reentry & pullout (T = 265 s, V = 4,500 km/h, H = 30,000 m)
6 – Approach phase (T = 450 s, V = 1,200 km/h, H = 2,600 m)
7 – Landing phase (T = 638 s, V = 151 km/h, H = 2,600 m)
8 – Touchdown (T = 684 s, V = 103 km/h)

Silver Lake

UTAH

Ely AFB

Beatty AFB

NEVADA

Edwards AFB

CALIFORNIA

a long, low spaceplane with fluid lines, a cruciform tail, and a wedge-shaped vertical fin; multiple rocket engines; a skin fashioned from the highly heat-resistant Inconel-X chrome-nickel alloy; and the capability to fly up to Mach 7, to reach several hundred thousand feet in altitude, and to be launched from a bomber/mothership.

The media flocked to Inglewood when North American unveiled the X-15 in October 1958, the world's first viewing of the sleek, black rocket plane. It looked like something borrowed from the future: long (49 feet/15 meters), tall (14 feet/4.3 meters at the vertical tail), heavy (31,275 pounds/14,186 kilograms launch weight), and powerful (57,000 pounds/25,855 kilograms of thrust from the XLR-99).

The X-15 represented a welcome testament to American technological know-how as it struggled to get its footing in the space race. It also helped to introduce the NACA's successor, the National Aeronautics and Space Administration (NASA), to the public, which opened its doors on the first day of October—in the same month as the X-15 rollout.

Scott Crossfield, who only five years earlier broke the Mach 2 barrier in the NACA's D-558 Skyrocket aircraft, piloted all the initial flights of the X-15 at NASA's High Speed Flight Station. His first flight almost ended in catastrophe; as Crossfield prepared for the drop from the B-52 mothership on June 8, 1959, he noticed that the pitch damper failed, but relying on the immense dry lakebed at Edwards in case of an emergency, he decided to go ahead. As he descended (without power) to the runway for a glide landing, strong longitudinal oscillations began. Relying on his extensive X-plane experience, Crossfield succeeded in landing at the bottom of one of the wild swings. He walked out uninjured, but X-15 number 1 took six months to repair. Crossfield piloted the *Black Bull* thirteen more times.

The subsequent X-15 flights held far more significance than setting high-speed and high-altitude records. During the long three and a half years between Sputnik 1 and the first launch of an American into space, the X-15 represented the United States at the dawn of this new form of exploration. By the time Alan Shepherd made his first suborbital flight in May 1961, the photogenic X-15 had flown thirty-six times and garnered media attention from around the world. Eventually, eight of its twelve pilots entered the astronaut corps, completing a total of thirteen missions at least 50 statute miles (264,000 feet/80,467 meters) above the surface of the Earth. And although he failed to qualify for this group, Neil Armstrong, the future commander of Apollo 11, did have the benefit of X-15 experience. He piloted it seven times from November 1960 to July 1962, flying as fast as Mach 5.74 and as high as 207,000 feet (63,094 meters).

But the X-15 did not only represent success; it also taught lessons about the risks of spaceflight. In November 1959, during the fifth X-15 flight, a small engine fire forced Scott Crossfield to make an emergency landing on Rosamund Dry Lake, breaking the spine of the aircraft, but avoiding harm to himself. In 1962, an engine failure forced NASA pilot Jack McKay to land at Mud Lake, Nevada, during which the landing gear collapsed, flipping the X-15 onto its back and inflicting career-ending injuries on McKay. Finally, in November 1967, Air Force Maj. Michael Adams died when his X-15 fell into a spin and then a dive before crashing.

In the end, the X-15 flight research program left a deep impression on human spaceflight. It flew 199 missions from June 1959 to October 1968. It claimed some remarkable performance milestones, going as fast as Mach 6.7 and as high as 354,200 feet (107,960 meters). It also served as one of aerospace history's most important testbeds, providing a trove of data for the manned space program about aeromedicine (investigating the physiological reaction of pilots experiencing vertigo and undergoing the rigors of spaceflight); hypersonic aerodynamics and thermodynamic heating (discovering unexpected drag at the tail section and hot spots on the protruding cockpit, as well as the front and lower sections of the aircraft); and hundreds of science platform experiments (covering, among many other subjects, the impact rate of micrometeorites and the extent of radiation from 70,000 to 100,000 feet/21,000 to 30,500 meters).

Perhaps most valuable of all, those in the succeeding generation who designed the space shuttle found in the X-15 data bank a treasury of real spaceflight experience from which to draw.

MiG 105-11

In 1966 the Soviets initiated the Spiral Project, based on a lifting-body concept much like the American HL-10. Three subscale models, known as BOR-1 to -3, flew from 1968 to 1969 at the Plesetsk Cosmodrome. Spiral ended when Soviet authorities decided to pursue the Buran shuttle—again, a parallel development to the American space shuttle whose origins dated to the 1960s. But before ending Spiral, between October 1976 and September 1978, the Soviets undertook a series of eight subsonic flight tests on a Soviet aircraft, the MiG 105-11. Its development started in 1965, ended temporarily in 1969, and resumed in 1974.

Known also as the experimental passenger orbital spacecraft (EPOS) program, the engineers who conceived of it took an imaginative approach, deciding not to design it as a classical lifting body like the HL-10 or Dyna-Soar, but instead to build it as a conventional delta wing with variable dihedral. At launch and reentry, its wings folded up 60 degrees; at subsonic speeds during approach, the pilot lowered the wings to horizontal,

Mikoyan-Gurevich (MiG): 105-11 1/1 EPOS Prototype
EPOS (Eksperimetalnaya Pilotiruemaya Orbitalnaya Samolyota)

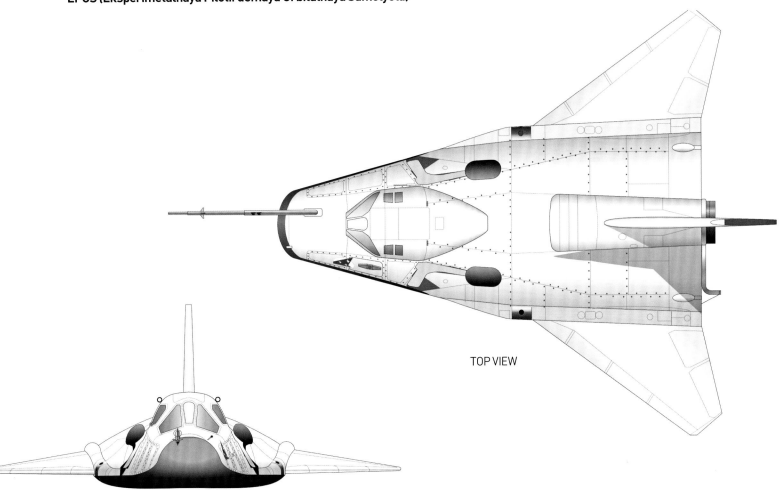

TOP VIEW

FRONT VIEW

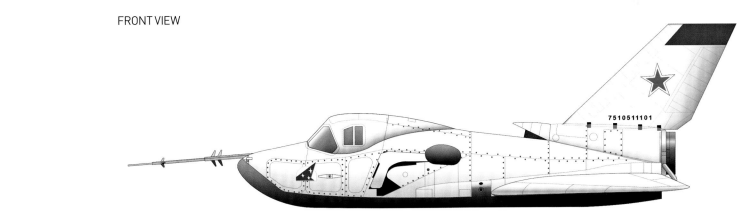

SIDE VIEW

0   1   2
meters

enabling better flight characteristics. Still, it bore a kinship to lifting bodies such as the HL-10, with its flat bottom and large, upturned nose (earning it the nickname of "the shoe"). The MiG weighed 9,300 pounds (4,220 kilograms) fueled and measured 34.9 feet (10.6 meters) long (including the instrument boom), just over 9 feet (2.8 meters) in diameter, with a wingspan of 22 feet (6.7 meters). Its propulsion consisted of one Koliesov RD-36-35K turbojet, as well as a rocket engine.

At least three pilots flight-tested the MiG 105-11: Gherman Titov, the second person to orbit the Earth, in Vostok 2; Vasily Lazarev, who later flew the first Soyuz 7K-T mission; and Aviard Fastovets, who piloted the majority of the atmospheric tests. The initial flight occurred on October 11, 1976, taking off from a dirt airstrip near Moscow and flying about 12 miles (19 kilometers) at an altitude up to 1,837 feet (560 meters). A year later, on November 27, the MiG made its first airdrop, launching from a Tu-95K bomber aircraft from over 16,400 feet (5,000 meters) to a landing on an airfield of packed earth. During one mission on an especially hot summer day, the aircraft sank into a black-topped runway. Project engineers responded ingeniously, freeing it by smashing watermelons on the airstrip and wedging watermelon halves under its skids.

Overall, the 105-11 flew under many conditions, with air-breathing and with rocket power, with and without skids, with ground takeoffs and air drops. The final flight in September 1978 ended in a hard landing, but those associated with the program felt that the eight missions represented an accurate assessment of its aerodynamics and propulsion.

Indeed, the design team at Mikoyan-Gurevich hoped to see the MiG 105-11 adapted directly for manned orbital flight. Instead, the configuration informed the design of a much larger and more ambitious spaceplane—one conceived to compete head-to-head with NASA's space shuttle.

LANDERS
LK

The Soviet parallel of the lunar module, or LM, became known as the LK (from the Russian "lunniy korabl," or lunar craft). As conceived, the LK weighed far too much; project engineers struggled all through the program to shave off kilograms; any engineer who cut just a kilogram from the scales received a 50–60 ruble prize. Ultimately, the LK carried only one cosmonaut (not two passengers, like the Apollo LM) to pare down the load. Another difference: no docking tunnel existed between the 7K-LOK and the LK, so once in lunar orbit, a cosmonaut needed to make a spacewalk to enter to the LK. And the American and

Soviet landers also parted company in their propulsion systems. The LM had one engine for landing and a separate one for lifting off from the moon; the LK combined both in one powerplant, but with ascent redundancy. The LK weighed one-third of the LM.

The LK looked like the LM's smaller brother. Its most prominent feature—a pressurized, semi-spherical lunar cabin—carried the cosmonaut on the trip to the lunar surface and back. The cabin rested on a circular rocket stage, and the entire structure sat on a four-legged landing gear. The hexagonal grid of the Kontakt docking system stood at the top of the spacecraft. Although shorter than the 23-foot (7-meter) LM, the LK still measured a considerable 17 feet (5.2 meters) in height, 7.4 feet (2.25 meters) in diameter, with a gross weight of 12,250 pounds (5,560 kilograms).

The Soviet space agency tested the LK in Earth orbit four times aboard Soyuz-L rockets, in November and December 1970, and in February and November 1971. In the end, engineers declared it fit for manned flight. But due to the repeated failure of the N1 rocket, the LK never saw service, and the victory of Apollo in the US ended the entire Soviet lunar program in May 1974.

Lunokhod Rovers 1 and 2

Although the Soviet space agency failed in its overall attempt to surpass the United States in the race to the moon, it did score many impressive victories. From 1959 to 1976, twenty-four Luna spacecraft reached Earth orbit on their way to the moon, including Luna 3 that captured the world's first image of its far side, Luna 9 that made the initial soft landing, and Luna 16 that returned surface samples.

One robotic Soviet mission to the moon stands out among the others. To support the upcoming manned expeditions, the Soviet space agency authorized the development of an automated lunar rover to assess the viability of the touchdown sites selected for the LK lunar landers. The rovers pursued two missions: to study the surface features for the later missions and to leave behind radio beacons to guide the landing of the LKs.

Design and fabrication of these spacecraft fell to the Lavochkin Design Bureau, under the personal guidance of Alexander Kemurdzhian, who worked on the project from beginning to end (1963 to 1973). After five years of research, the rovers underwent testing outside of a secret village near Simferopol, on the Crimean Peninsula, where crews erected a simulated moonscape roughly 400 feet (120 meters) by 330 feet (70 meters) with fifty-four craters and 106,000 cubic feet (3,000 cubic meters) of soil. Here, technicians drove the rovers over the simulated lunar terrain to learn their capacities and idiosyncrasies.

LK Lunar Lander (11F94)

TOP VIEW

TOP SECTION

FRONT VIEW

FRONT SECTION

SIDE VIEW

SIDE SECTION

0 1 2 meters

These robotic machines, called lunokhods ("moonwalkers"), received launch support from the Proton rocket and traveled to the moon on board the Luna spacecraft. The lunokhods' appearance suggested something functional and durable: a deep tub covered by a hinged, convex lid, with four large, independently powered wheels on each side. But its simple looks belied its clever design—it had extendable probes to test the moon's soil for density and structure, two low-resolution television cameras, four high-resolution photometers, as well as an x-ray spectrometer, an x-ray telescope, and radiation detectors. The rover used battery power, recharged during sunlit hours by a solar array on the inside of the lid. Relatively compact and light, but certainly not small or delicate, it weighed 1,667 pounds (about 756 kilograms) without fuel and measured about 4.4 feet (1.35 meters) tall, just over 7 feet (2.15 meters) long, with a wheelbase of about 5.2 feet (1.6 meters).

On its initial launch, the project suffered an embarrassing setback. Lunokhod 201 lifted off on February 19, 1969, but within seconds the Proton rocket disintegrated, destroying the payload. In the process, polonium 210, used to heat the rover's instruments at night, scattered over a wide area.

The next test came on the Luna 17 mission, when the Lunokhod 1 entered the moon's orbit on November 15, 1969. At 9.9 miles (16 kilometers) altitude, the Luna's braking rockets ignited, followed by the firing of the main thrusters, and then the secondary thrusters. At 5 feet (1.5 meters) above the moon's surface, the engine cut off and the spacecraft soft-landed in the Sea of Rains. Dual ramps on the Luna mother ship enabled the Lunokhod to roll out onto the moon's surface. There, from November 17, 1970, until September 14, 1971—298 Earth days—drivers at mission control directed the rover on a journey of 6.55 miles (10.54 kilometers), moving about by day and hibernating at night. It captured twenty thousand television images and 206 high-resolution panoramas, tested twenty-five soil samples using the on-board x-ray fluorescence spectrometer, and deployed its penetrometer at 500 points.

Lunokhod 2 survived significantly fewer days than its brother spacecraft but far outdistanced it. During the Luna 21 mission, it landed in the Le Monnier Crater on January 18, 1973, and traveled until June 3—136 days in all. It failed when dust fell into the open lid and dropped onto the spacecraft's radiators when the lid closed, causing it to overheat and become inoperable. Before that point, it traversed 24 miles (39 kilometers), took eighty thousand television images and eighty-six high-resolution panoramas, and conducted mechanical tests of the surface, laser ranging measurements, and other experiments.

For years, the Lunokhods stood alone as the only rovers ever to land on another celestial body. Only in 1997 with NASA's Mars Pathfinder project (and its Sojourner rover) did another space

vehicle achieve that feat. And not until *July 2014* did the Mars exploration rover Opportunity surpass the distance record held for forty-one years by Lunokhod 2.

STATIONS
Skylab

Even before the first lunar touchdown of Apollo 11 in July 1969, NASA officials looked ahead to the post-Apollo era. Starting in 1968, they considered many potential projects that could follow the moon missions and called these prospects the Apollo Applications program. They envisioned the program as a way to repurpose some of the leftover Apollo hardware and use it to send astronauts into space on science missions. Among these projects, a space station concept gained traction, in part because as Apollo Applications took shape, the USSR prepared for a series of stations—the first of which, Salyut 1, went into orbit in 1971.

Refashioning Apollo components into a viable space station required considerable ingenuity. Its engineers envisioned it as a long string of connected modules, disappointingly unlike the majestic wheels popularized by artist Chesley Bonestell in the 1950s. The final design looked like a long, progressively widening tube decorated with massive solar arrays.

When crews arrived at the completed station, they left the Apollo command and service module on which they arrived and entered the multiple docking adapter. At 90 degrees to the adapter, an octagonal science laboratory called the Apollo telescope mount jutted out. This observatory enabled the astronauts to make sightings of the sun and other space phenomena on nine different astronomical instruments. Going straight ahead through the docking adapter, the astronauts reached the airlock module, where they could embark on spacewalks. After crossing a narrow, ring-shaped component known as the instrument unit, they reached a long, wide cylinder, the orbital workshop. This was their main living and working quarters with laboratory space. The station had two sources of power: the telescope mount drew electricity from four solar arrays on its exterior and the station got it from two panels, attached to the orbital workshop.

Among the six main components of this spacecraft, the command and service module, the instrument unit, and the orbital workshop all originated in the Apollo program—respectively, as the astronaut capsule, as the Saturn V's storage unit for guidance equipment and computers, and as the giant rocket's hydrogen fuel tank. From end to end, this station, called Skylab, measured more than 120 feet (almost 37 meters) long and weighed nearly 170,000 pounds (over 77,000 kilograms).

Apollo Applications Program (AAP)
Skylab 1

0 1 2 3 4 5 meters

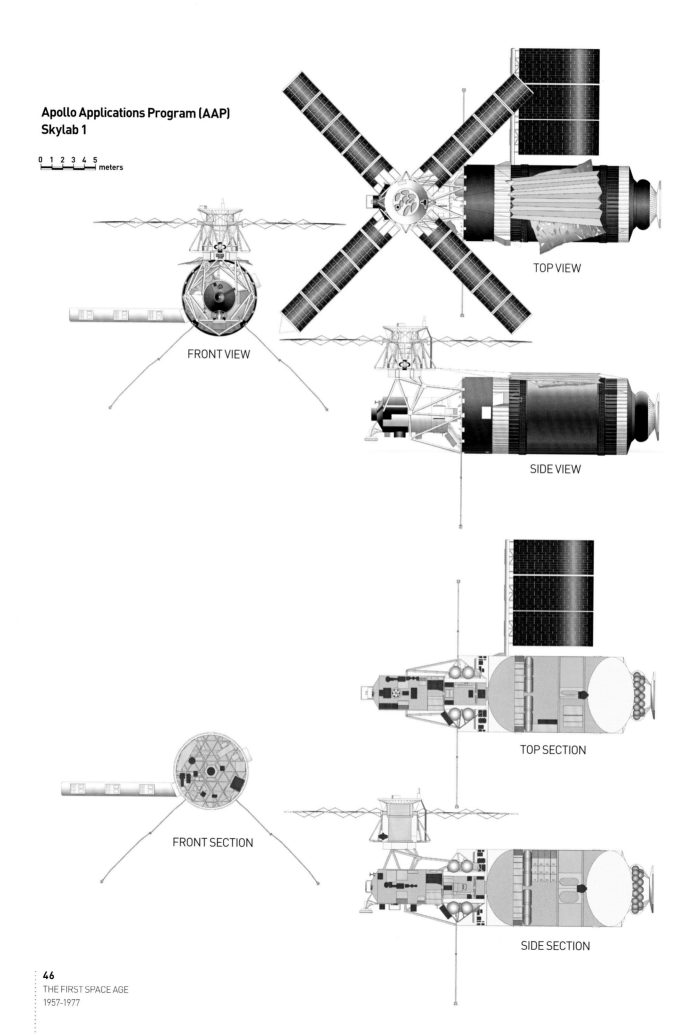

TOP VIEW

FRONT VIEW

SIDE VIEW

TOP SECTION

FRONT SECTION

SIDE SECTION

Skylab entered space on four separate boosters. The Saturn V—configured in two stages on this occasion to allow for the big payload shroud that protected the spacecraft—launched the station itself into orbit on May 14, 1973, with the name of Skylab 1. Subsequently, Saturn 1B rockets lifted three crews of three astronauts each during Skylab 2 on May 25, Skylab 3 on July 28, and Skylab 4 on November 16, 1973.

Skylab operations began with a high likelihood of outright failure. As Skylab 1 rose from the launch pad, a micrometeoroid shield designed to protect the spacecraft broke loose from its mountings, striking and jarring open one of the two main solar arrays. By the time Skylab reached orbit, both parts had blown off. Meanwhile, the other main array became tangled in the ensuing debris and failed to deploy. Without the shield, temperatures inside Skylab rose to 126 degrees Fahrenheit (52 degrees Celsius).

NASA tried to salvage Skylab. Instead of launching Skylab 2 the day after Skylab 1, as planned, NASA delayed the flight for two weeks as its engineers considered repair options and the astronauts practiced in Houston for spacewalk scenarios. In the end, after intensive trial and improvisation, the Skylab 2 astronauts (Charles Conrad, Paul Weitz, and Joseph Kerwin) arrived at Skylab with a solution. They deployed a 22-foot (6.7-meter)-by-24-foot (7.3-meter) parasol that blocked the sunlight and lowered temperatures in the cabin. They also managed to release the jammed solar panel. By the end of their twenty-eight-day mission, they completed about 80 percent of their scheduled solar observations.

Skylab 3 also experienced difficulties. The least threatening involved the three astronauts (Alan Bean, Jack Lousma, and Owen Garriott), all of whom fell prey to nausea for about a week. Despite their illness, they augmented the parasol installed during Skylab 2 with a two-post sunshade, requiring two of the crew to undertake a record EVA lasting more than six and a half hours. More urgently, they reported leaks in two thrusters on the command module, and NASA engineers made plans either for an emergency return to Earth or for a rescue mission. Another EVA failed to find the source of the leaks. Ultimately, ground staff worked out special procedures for a safe flight home, despite the problems. Skylab 3 ended after fifty-nine days.

Skylab 4 went more smoothly, but sickness recurred among the crew. When Gerald Carr, William Pogue, and Edward Gibson recovered, they conducted a six-hour, thirty-four-minute EVA to repair a jammed antenna. Scientifically, Skylab 4 contributed unique observations of the famed comet Kohoutek. The astronauts did EVAs on Christmas Day and on December 29, 1973, to take photographs of the comet as it passed. Skylab's astronauts remained on station for eighty-four days, nearly as long as the first two missions combined.

Skylab gave NASA a foretaste of space station life. In the combined 171 days aloft, the nine astronauts undertook major, frequent, and often dangerous EVAs to repair the station, conducted complex astronomical viewings, and acted as physiological subjects testing the effects of long-term exposure to spaceflight. The crew also experienced the personal peculiarities of long-term habitation, involving such practicalities as sleep, exercise (they used a portable treadmill), food, and waste management.

When the time came twenty-five years later to begin assembling the International Space Station—and during the planning and fabrication process leading up to it—the Skylab experiences served as an indispensable manual by which to anticipate the unforeseen.

Apollo-Soyuz Test Project

If President John F. Kennedy used the US space program as a "soft" weapon in the Cold War, during the 1970s his political nemesis Richard Nixon wielded it to deescalate hostilities with the Soviets. The process began early in 1972, when President Nixon made a historic trip to China. Nixon calculated correctly that the Soviets would view a warming US-China relationship with alarm, so the following May he became the first president to visit the USSR. Nixon not only made agreements related to arms control, cooperative scientific research, and expanded trade, but on May 24 he signed a commitment for the peaceful exploration of outer space. This pact later became a joint mission, Apollo-Soyuz test project (ASTP).

In addition to its contribution to international diplomacy, ASTP also served a domestic purpose: the Apollo program had almost run its course by this point. So Apollo-Soyuz injected some anticipation and excitement into a project already losing much of its appeal due to repetition. A good deal of surplus Saturn and Apollo hardware still remained in storage, ready for reuse. For its part, the Soviet space program—buffeted by internal problems as well as the triumph of Apollo—needed the lift that this goodwill flight offered.

Originally, the two sides discussed a simple rendezvous of the Apollo and Soyuz spacecraft at the Salyut 4 space station. But perhaps fearing the appearance of an imbalance between the Soviet and American contributions, they instead agreed to a docking between the Soyuz and Apollo capsules. This plan required each to design its own module, as well as to collaborate on a shared airlock module. The resulting Apollo-Soyuz spacecraft and docking module measured about 71 feet (21.65 meters) long and weighed approximately 51,964 pounds

Apollo-Soyuz Test Program
Flight Configuration

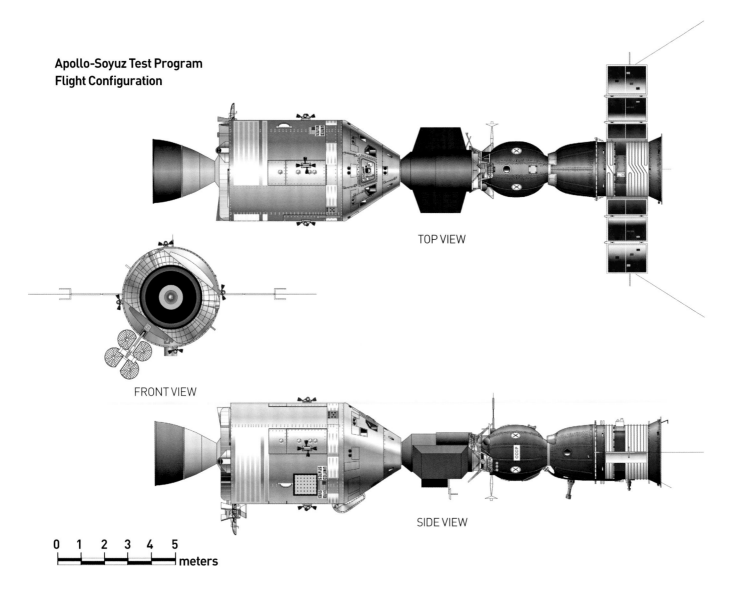

TOP VIEW

FRONT VIEW

SIDE VIEW

0 1 2 3 4 5 meters

Salyut 1 (DOS-7K #1)

FRONT VIEW

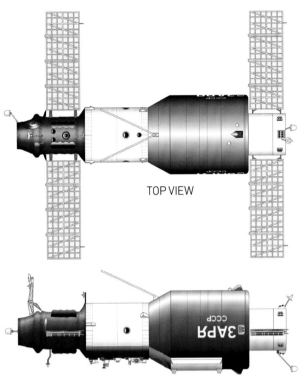

TOP VIEW

SIDE VIEW

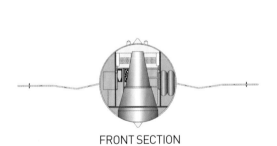

FRONT SECTION

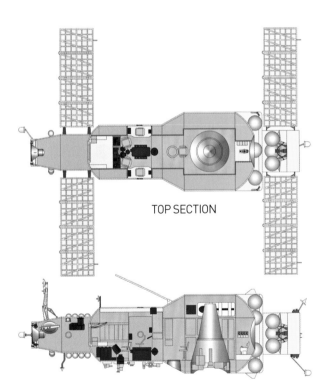

TOP SECTION

SIDE SECTION

0 1 2 3 4 5 meters

Salyut 4 (DOS-7K #4)

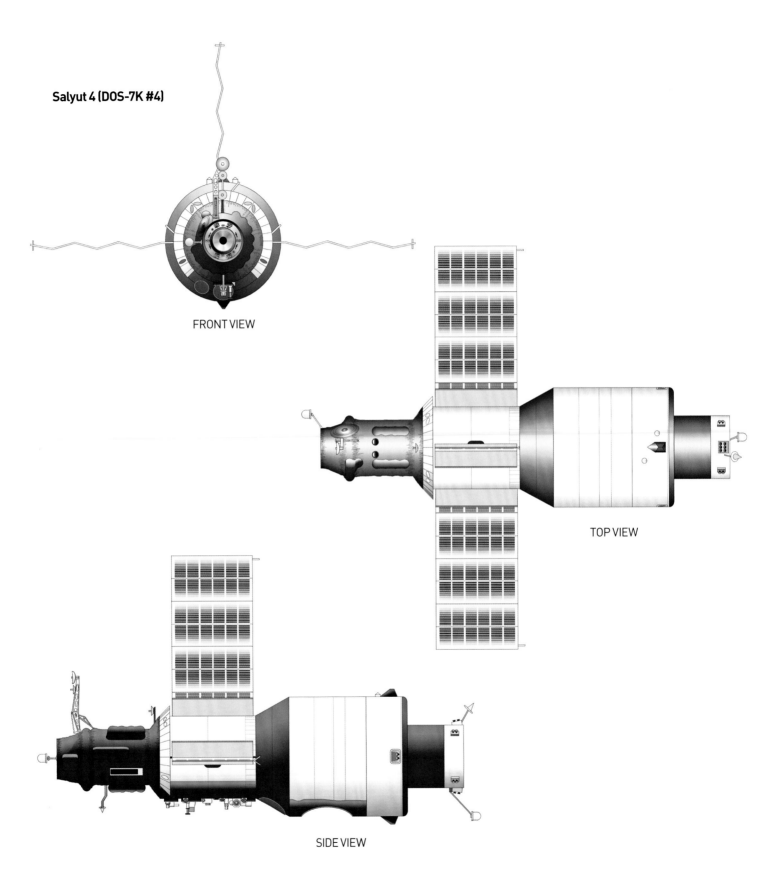

FRONT VIEW

TOP VIEW

SIDE VIEW

0 1 2 3 4 5 meters

Salyut 6 (DOS-7K #5)

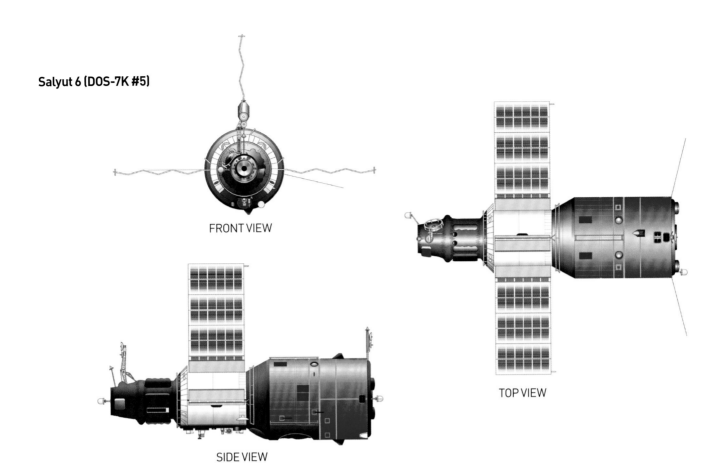

FRONT VIEW

TOP VIEW

SIDE VIEW

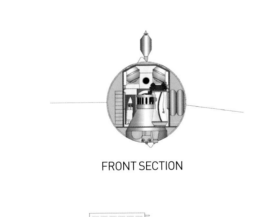

FRONT SECTION

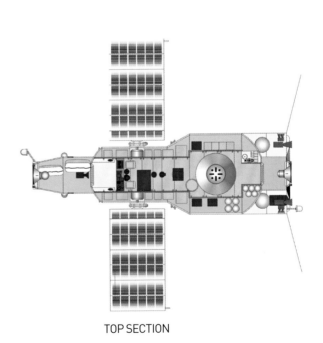

TOP SECTION

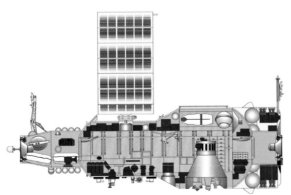

SIDE SECTION

0 1 2 3 4 5
meters

Almaz OPS (Salyut 2, 3 & 5)

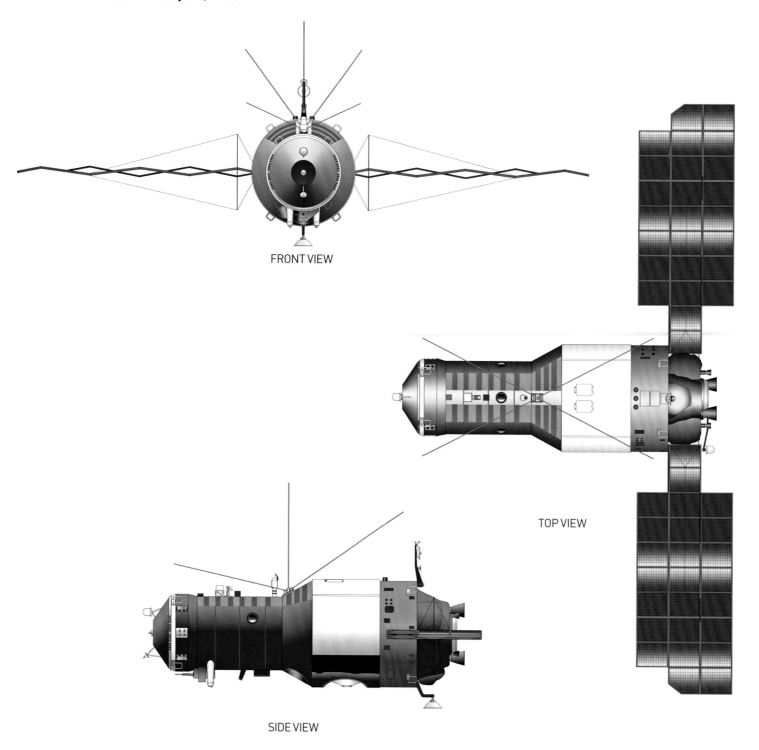

FRONT VIEW

TOP VIEW

SIDE VIEW

0 1 2 3 4 5 meters

(23,570 meters)—certainly not a small or light vehicle by the standards of the day.

On the appointed launch date of July 15, 1975, the Soviet and American sides succeeded in a complex liftoff and docking sequence. Apollo took off from the Kennedy Space Center aboard the Saturn IB (the last of thirty-two made); seven and a half hours earlier, the Soyuz 7K-M capsule left the Baikonur Cosmodrome in Kazakhstan on the Soyuz-U rocket. At Soyuz's thirty-sixth and Apollo's twenty-ninth orbit, the spacecraft approached one another and Apollo initiated active docking maneuvers. They mated smoothly, six minutes ahead of schedule. Three hours later, the Soyuz hatch opened and the astronauts and cosmonauts came forward and shook hands.

For the Americans, the crew consisted of Commander Tom Stafford, Docking Module Pilot Deke Slayton (of Mercury fame, now fifty-one years old), and Command Module Pilot Vance Brand. The Soviets sent Commander Alexei Leonov (the first man to walk in space) and Flight Engineer Valeri Kubasov. Both sides came equipped with at least a basic knowledge of the other's language, enabling them to work together on joint experiments in the docking module, to eat together, to visit each other's capsules, and to hold a joint televised news conference.

After forty-three hours, the crews parted. The Soviets returned to Kazakhstan after almost six days in space; the Americans stayed aloft doing Earth observation experiments for another three days before a live press conference and then a splashdown in the Atlantic.

The Apollo-Soyuz Test Project represented a small but decisive step toward ending the almost twenty-year competition between America and the USSR in space. Because of the close cooperation necessary for ASTP to succeed, it set an important precedent, redeemed in the 1990s with the planning and construction of the International Space Station.

Salyut 1

Even before the Apollo 11 lunar landing in July 1969, Soviet space agency officials realized that their efforts to reach the moon before NASA had a low probability of success. The USSR's indispensable N1 rocket had failed twice by this time; two more bad launches in 1971 and 1972 doomed it to oblivion. Having almost certainly lost the moon race, the Soviets anticipated further embarrassment from NASA's plan to launch the world's first space station, known as Skylab. The Soviets decided that to regain the initiative, they needed to claim the long-duration flight mission for themselves, which meant beating Skylab to the launch pad. Luckily, they already had a station in progress. It was the Almaz military

spacecraft, then under fabrication at the Vladimir Chelomei design bureau. It had originated as a Soviet response to the US Air Force's manned orbiting laboratory (MOL), pursued from 1963 to 1969. Even more fortunate, the Proton rocket, another of the Chelomei group's designs, represented a reliable launch option for the Soviet space station.

To expedite the project, Chelomei's engineers proposed a collaboration with the design bureau of Visily Mishin. Mishin disapproved of the partnership with Chelomei, wishing instead to build a station proposed by Korolev before his death. But the pressure to head off Skylab—as well as the camouflage that the civilian Soyuz offered to the military Almaz—overcame these objections. Representatives from both groups met and agreed to adapt the Almaz for civil space purposes on an eighteen-month timeline, starting in February 1970. The project engineers foresaw three main tasks: building an exit at the forward end of Almaz using a Soyuz docking system and airlock; installing an engine borrowed from Soyuz at the aft end; and mounting Soyuz solar panels at these new sections.

A year later, the civilian Salyut (Salute) 1 space station left the Chelomei facilities on its way to the Baikonur launch complex in Kazakhstan, where it underwent final assembly. Not especially big or heavy by twenty-first century standards, Salyut 1 nonetheless seemed like a behemoth in its time. It weighed about 41,667 pounds (18,900 kilograms) at launch and measured almost 52 feet (15.8 meters) long, with a maximum diameter of 13.6 feet (4.15 meters) and a span of nearly 33 feet (10 meters) at the solar arrays. The interior space of Salyut 1 differentiated it from every other space vehicle until then; its cosmonauts had at their disposal an unprecedented 3,178 cubic feet (90 cubic meters) of interior room. It consisted of three segments: a transfer compartment with the spacecraft's sole docking port at the forward end; a main, cylindrical portion containing space for up to eight work stations; and auxiliary sections holding the communications and control equipment, power supply, and life support system. Salyut 1 also carried a mirror telescope, the first ever to make observations from space.

With the launch of the first Skylab still pending, Salyut 1 lifted off on April 19, 1971. During that year, two crews visited, both transported by the new Soyuz 7K-OKS capsule. On the first mission (Salyut 10, on April 23, 1971), the crew soft-docked with the station but failed to connect solidly due a design flaw in the Soyuz's autopilot. Soyuz 11 followed and succeeded in berthing with Salyut on June 7. Its crew of Georgy Dobrovolvsky, Viktor Patsayev, and Vladislav Volkov remained for almost twenty-four days. Their return home seemed normal until technicians opened the hatch after it landed; they found all three men dead.

A subsequent investigation determined that a pressure relief valve failed, causing a loss of cabin atmosphere.

Meanwhile, Salyut 1 remained in orbit for 175 days, although it welcomed no more cosmonauts. Understandably, the Soviets retired the Soyuz 7K-OKS after the loss of the three crewmembers, and its successor, the Soyuz 7K-T, needed to undergo flight testing before being cleared for spaceflight. The Soviet space station program returned to flight with Salyut 2 on April 3, 1973—just seven weeks before Skylab made its maiden flight.

Salyut 2, 3, 5 (Almaz)

At the end of 1969, Soviet military authorities accelerated the secret Almaz program, an implicit admission that the Americans had won the race to the moon. By January 1973, the first Almaz (referred to publicly as Salyut 2) arrived at the Baikonur Cosmodrome.

At its forward end, Almaz consisted of a main cylindrical compartment for a crew of up to three; at its midsection, it had an auxiliary hold for life support, communications and control equipment, and the power supply; and aft, a docking port with two solar arrays. It weighed 41,780 pounds (18,950 kilograms) at launch and measured 47.7 feet (14.55 meters) long with a diameter of 13.6 feet (4.15 meters). Its cosmonauts lived in 3,178 cubic feet (90 cubic meters) of habitable space. The civilian Salyuts differed from Almaz in having four (rather than two) solar arrays, a forward transfer compartment, and in measuring an extra 4 feet in length.

Almaz and her crews flew during a four-year period, from spring 1973 to summer 1977. But this timespan gives a false impression of the endurance of each mission.

The first Almaz (Salyut 2) proved to be ill-fated. Ten days after its April 3, 1973, launch, the flight plan called for it to rendezvous and dock with a Soyuz capsule carrying Commander Pavel Popovich and Flight Engineer Yuri Artyukhin. But the Soyuz encountered technical problems, grounding it. In the meantime, after Salyut 2 went into orbit, controllers noticed decreased air pressure on the station. Subsequent investigations showed that the Proton rocket's upper stage exploded three days after liftoff, raining debris onto the station. Eight days later, one of these pieces struck it, causing the depressurization.

The second, Salyut 3, fared better. It arrived on orbit on June 25, 1974, carrying a payload consisting mostly of surveillance cameras: the Agat-1 with a resolution of more than 3 meters, as well as panoramic, topographical, star, and Volga infrared equipment. It also introduced a self-defense gun mounted at the forward part of the station. Flying as the Soyuz

14 mission, its crew—Popovich and Artyukhin, held over from Salyut 2—traveled to the station on July 3 and docked manually on July 4. The cosmonauts activated the cameras, took photos of Central Asia, and undertook minor chores to make the station more comfortable. Soyuz 14 ended on July 19 when the crew returned to Earth. Soyuz 15 attempted to visit the station on August 26, 1974, but problems with the rendezvous system forced the cosmonauts to return home. Salyut 3 burned up in the atmosphere on January 24, 1975.

The last of the Almaz spacecraft, Salyut 5, entered service on June 22, 1976. Three missions visited it: Soyuz 21 (July 6 to August 24, 1976), Soyuz. 23 (October 14–15, 1976), and Soyuz 24 (February 7–25, 1977). The first one, manned by Commander Boris Volynov and flight engineer Vitaly Zholobov, had to be curtailed a week before the end of its scheduled two-month stay due to a bad smell in the cabin and headaches among the crew. The second mission failed to dock due to a malfunction in the automated system. Soyuz 21 encountered a similar incident, but succeeded in docking manually. Finally, the last Almaz crew, Viktor Gorbatko and Yuri Glazkov, replaced the air contaminated during the previous mission, conducted some solar experiments, but cut the mission short because of a depletion in the station's main engine propellant. Salyut 5 de-orbited on August 8, 1977.

If nothing else, the three Almaz spacecraft taught an invaluable lesson: that several stubborn obstacles remained to be overcome before the Soviet space agency could claim that it achieved safe, dependable, long-duration habitation in space.

Salyut 4

Among the first five Salyut/Almaz space stations launched by the USSR, only Salyut 4 showed the promise of future long-duration spaceflight. The basic structure of Salyut 4 differed only slightly from that of Salyut 1. Instead of the earlier spacecraft's small-sized solar arrays, Salyut 4 generated power with three larger panels on the forward module. Additionally, it carried a 4,409-pound (2,000-kilogram) instrument suite, including the orbiting solar telescope with a 10-inch (25-centimeter) diameter mirror; two x-ray telescopes; a spectrometer for far ultraviolet observations; and x-ray detectors and optical sensors installed on the exterior of the station.

Launched on December 26, 1974, Salyut 4 hosted three missions. Soyuz 17 arrived on January 12, 1975 with Commander Aleksei Gubarev and Flight Engineer Georgy Grechko docking manually to the station. When they entered Salyut for the first time, they found a comedic note left by someone in the factory, telling them to "Wipe your feet!" The crew worked fifteen-to

Almaz OPS (Salyut 2, 3 & 5)

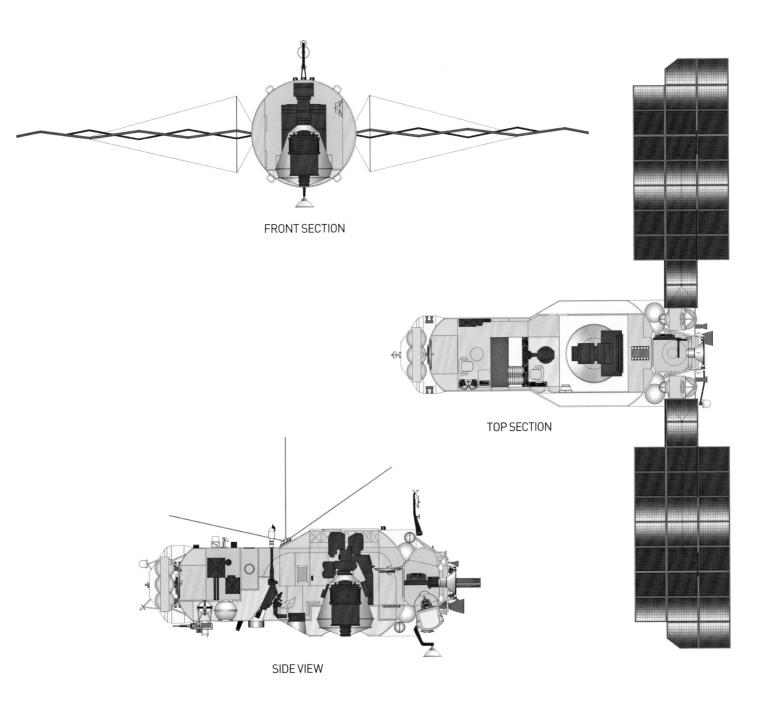

FRONT SECTION

TOP SECTION

SIDE VIEW

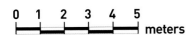

0 1 2 3 4 5 meters

twenty-hour days, testing communications equipment used to track ship movements, and they also resurfaced the mirror on the station's solar telescope. It had been damaged by direct exposure to the sun when its pointing system failed. After twenty-nine days on board, they returned to Earth on February 9.

After aborting Soyuz 18 because of the failure of its Soyuz launch vehicle, Soyuz 18 Commander Pyotr Klimuk and Flight Engineer Vitaly Sevastyanov docked with the station on May 26, 1975. They conducted a wide variety of tasks, some related to housekeeping and others involving biological research, medical experiments, and astronomical observations of the Earth, other planets, and stars. In July, the spacecraft's environmental system began to fail, causing mold to grow on the walls. The cosmonauts returned to Earth on July 26, 1975, after almost sixty-three days on the station, a record for Salyut to date.

During the Soyuz 20 mission, the Soviet space program made more progress, but without a crew. The Soyuz 7K-T spacecraft launched on November 17, 1975, and docked at Salyut 4 for ninety days, doing automated checks of upgraded systems and conducting a biological experiment that exposed plants to the spacecraft's climate for three months. The 7K-T departed on February 16, 1976.

After 770 days in orbit (92 occupied), Salyut 4 burned up as it reentered the atmosphere on February 3, 1977. While not yet the high-water mark of Soviet space station endeavors, Salyut 4 marked a solid achievement on which to build future preeminence, realized in the later Salyut stations and finally in Mir.

ROCKETS
Vanguard 1

Choosing a civilian instead of a military satellite to represent America's first foray into space looked at the time like a sensible decision. The Eisenhower administration did not want to exploit the obvious candidates of the armed forces—the air force's Atlas, the army's Redstone, or the navy's Polaris ballistic missile—because time spent developing any of them for peaceful purposes might delay their deployment against the Soviet Union. In addition, it conformed to President Eisenhower's belief that space exploration should be a civilian pursuit, in part to obscure the government's parallel, covert program of space espionage. So when the organizers of the International Geophysical Year (IGY) announced in 1955 a competition to develop satellites capable of scientific observations of the Earth, the administration chose Project Vanguard, sponsored by the Naval Research Laboratory Project (NRL).

Vanguard also seemed to be a logical choice because of the NRL's ongoing and successful Viking rocket program. Although a part of the Department of the Navy, the laboratory and a group of its project engineers conceived of Viking in 1945 not as a weapon, but as a vehicle for researchers to conduct science experiments and Earth observations. Fabricated by the Martin Company and Reaction Motors, it looked superficially like the V-2, but weighed two-thirds less, carried a 100-pound (45-kilogram) payload, and attained a similar altitude. Viking underwent twelve firings from 1949 to 1955, reaching 50 miles (80 kilometers) at the beginning and 158 miles (254 kilometers) at the end—a record altitude for a single-stage rocket. Its missions became the first to measure upper atmosphere winds, temperature, and pressure; the first to take high-altitude pictures of the Earth; and the first to photograph hurricanes and tropical storms. Moreover, Viking's engine gimballing system pioneered steering on rockets.

In winning the Vanguard project over the other competitors, the NRL made Dr. John Hagen its leader and Milton Rosen its technical director. Rosen envisioned the Vanguard rocket's first stage as a stretched version of Viking, powered by the General Electric liquid propellant engine used in the US Army's defunct Hermes missile project. He picked the navy's Aerobee missile, paired with an Aerojet liquid propellant engine, as stage two. Project managers subsequently added a third stage with a solid propellant engine, above which the spherical Vanguard satellite perched on a shaft. Viking's record of effectiveness; Rosen's preference for familiar, well-tried constituents; and the opportunity to pursue it in a civilian context all affirmed the Eisenhower administration's choice of Vanguard to answer the IGY call.

But two years after the decision, the NRL's candidate began to look less appealing. The advantage of using off-the-shelf components championed by Rosen diminished when detailed engineering analyses showed the need for extensive modifications. And the original $12 million budget swelled to $100 million. However, three early test flights gave reason for optimism. Launched on December 8, 1956; May 1, 1957; and October 23, 1957, each one succeeded—but they only fired the first (Viking) stage; none tested the entire, mission-ready stack. This trial came on December 6, 1957, when the complete vehicle plus the full guidance and control system waited on the launch pad at Cape Canaveral. It measured 75 feet (23 meters) tall, almost 4 feet (1.14 meters) in diameter, and weighed 22,156 pounds (10,050 kilograms).

The countdown brought disaster. At liftoff, Vanguard rose a few feet, stopped when its first stage lost power, sank back to the ground, and blew up in an enormous fireball.

This spectacular public failure not only received intense media coverage but came close on the heels of the historic successes of Sputniks 1 and 2 (on October 4, and November 3, 1957, respectively). These events persuaded the Eisenhower

Mercury-Atlas D

0 1 2 3 4 5 meters

Juno I–Explorer 1

Mercury–Redstone

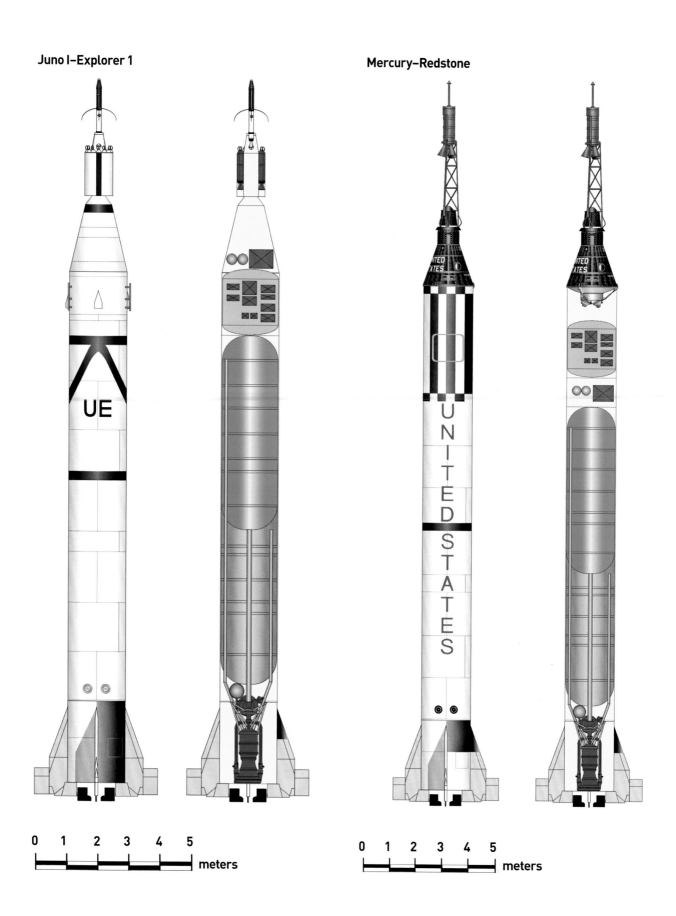

0 1 2 3 4 5 meters

0 1 2 3 4 5 meters

administration to bring back to life one of Vanguard's previous rivals, the army's Redstone (allied with the Jet Propulsion Laboratory's Sergeant missile), first proposed in 1955. The army/JPL team proved on January 31, 1958, to be the one that launched America into space with the Explorer 1 satellite.

Still, the NRL's candidate won vindication on March 17, 1958, when the Vanguard 1 satellite went into orbit. And despite its second-place finish, Vanguard left a deep imprint on American rocketry: its second stage wound up on the Thor missile, its third stage migrated to the Atlas missile, and its gimballing motors became part of the Saturn rocket design.

Redstone Juno 1

In contrast to the less-mature Vanguard program, the army had at its disposal a missile under development since the early 1950s, led by the German émigré Dr. Wernher von Braun and his coworkers at Huntsville (in association with the Chrysler Corporation). Designed originally as an intermediate-range nuclear missile capable of carrying a 6,900-pound warhead about 180 miles, the Redstone (named for the army's Redstone Arsenal at Huntsville) enjoyed the distinct advantage of employing many of the same engineers and scientists who made the V-2 rockets during World War II; indeed, the Redstone descended directly from the V-2. During thirty-seven test flights of the Redstone from 1953 to 1958, 65 percent flew successfully, but more importantly, during this five-year period von Braun's engineers made many important incremental improvements.

As the Redstone progressed in Alabama, the rocketeers at the Jet Propulsion Laboratory fashioned the smaller, single-stage, solid-propellant Sergeant tactical nuclear missile. Begun in 1955, the missile launched in 1956.

The Redstone and Sergeant missile teams joined forces in accordance with the president's go-ahead in December 1957, with a mandate to launch a satellite into orbit in just eighty days. Huntsville got the assignment to build the first stage; JPL agreed to make stages two to four, a communications system, and the satellite itself. The main task at Pasadena required the engineers to scale down the size of the Sergeant's solid-propellant rocket engines and, in stages 2 and 3, cluster eleven and three of them, respectively, in an outer-ring pattern. A single Sergeant engine constituted the fourth stage, which boosted the payload into space. To convert the Redstone from a warhead carrier to a satellite launcher, the von Braun group lengthened the missile by 8 feet for added tankage, strengthened the nose section to support the weight of the upper stages, and changed the fuel mixture. This revamped missile got the name Jupiter-C, which included all but the JPL fourth stage.

By adding the top stage and the payload to the Jupiter-C, the combined stack became known as the Juno 1, the rocket that boosted America's first satellites into orbit. Small by twenty-first century standards at 70 feet (21.2 meters) tall, almost 6 feet (1.8 meters) in diameter, and weighing just over 64,000 pounds (29,060 kilograms), it put three of its six payloads into orbit. But the success of its first one—the Explorer 1 spacecraft that lifted off on January 31, 1958—caused relief and celebration in America and restored some sense of optimism after the shock of the Soviet successes in Sputniks 1 and 2.

Mercury-Atlas

During the initial phase of Project Mercury in the summer of 1958, the space task group of the National Advisory Committee for Aeronautics approved a suggestion from one of its engineers for an off-the-shelf missile to boost the first Americans into orbit. Spacecraft designer Maxime Faget's proposal represented an obvious savings not only in time, but in money. But Faget's plan also underestimated the difficulty of converting a military weapon for civilian use—a weapon still in its developmental stage and one on which human life depended.

The evolution of the Atlas missile recommended by Faget began in January 1951 when the US Air Force (USAF) awarded a contract to Convair Aircraft to study the feasibility of an intercontinental ballistic missile (ICBM) capable of carrying an 8,000-pound nuclear warhead 5,000 miles to a target. Two years later, the requirements changed with the advent of the hydrogen bomb. The warhead associated with this new, thermonuclear explosive weighed significantly less than that of the atomic bomb, reducing the required payload capacity of Atlas, and making accuracy less important due to its higher yield. In response to these changed circumstances and to the onrush of Soviet missile technology, in 1954 the Defense Department accelerated Atlas, giving it the highest national priority.

Atlas posed a steep technological climb, requiring three main innovations: swiveling engines, a nose cone/warhead that separated from the missile before reaching its target, and, most radical of all, an all-new system of internal bracing for the missile itself. Instead of relying on the heavy structures that gave strength to earlier rockets, an engineer at Convair, Charlie Bossart, conceived of a system in which thin-walled monocoque propellant tanks filled with pressurized nitrogen (jokingly referred to as "steel balloons") served as the backbone of Atlas—essentially using vapors as the main weight-bearing component. Bossart's technique simplified construction and reduced the mass of the vehicle.

Flight testing of Atlas A began in 1957. The first two launches failed, but the third, in December, succeeded. In all, Atlas A met the USAF's expectations in only in three of eight tries. Atlas B did better, with five satisfactory flights (the first in August 1958) and three accidents. Three of six Atlas Cs flew as expected. Atlas D—the first model to be deployed to silos—experienced problems in its initial flight in April 1959, but it improved steadily, producing 90 satisfactory outcomes from 117 test launches.

NASA found itself integrating the Mercury capsule with the Atlas D amidst this uncertain, teething stage of the rocket's development. Despite the headaches, Atlas looked impressive on the launch pad at 82 feet (25 meters) long, 10 feet (3.05 meters) in diameter, with a weight of 255,900 pounds (116,100 kilograms).

Mercury-Atlas 1—the first attempt to launch the capsule and the booster together—did not raise anyone's confidence. The flight on July 29, 1960, ended after a little more than three minutes due to buckling of the rocket's skin just below the capsule. Convair and NASA engineers resolved the problem by reinforcing the booster/spacecraft interface with an 8-inch steel band and by stiffening the adapter that connected them. Following these steps, Mercury-Atlas 2 in February 1961 succeeded, and encouraged by this outcome, project managers scheduled Mercury-Atlas 3 for April 25, 1961. This was to be the first unmanned orbital mission. But it failed too, just thirteen days after Yuri Gagarin became the first person to orbit the Earth.

Before strapping in an astronaut, NASA made one last test flight: Mercury-Atlas 5 in November 1961 launched a 39-pound chimpanzee named Enos into space. The rocket performed well, its separation from the spacecraft occurred in sequence, environmental controls and tracking systems worked as planned, and Enos survived two orbits and 181 minutes of weightlessness in good physical condition.

So, although the Mercury-Atlas rocket began its career raising justifiable worries in the minds of NASA's leaders, in the end it provided reliable launch services that put the US space agency in sight of, if not yet in range of, its Soviet rival.

Agena

Named for one of the brightest stars in the night sky, the Agena upper-stage rocket became an indispensable part of NASA's space missions. And like Centaur, Agena originated not with the US space agency, but with the USAF.

Two years before the launch of Sputnik 1 in 1955, the USAF initiated Weapon System 117L, an ingenious hybrid later known as Agena. On the one hand, it served as a second-stage that lifted satellites into orbit. But its designers also envisioned it as a photo-reconnaissance satellite, built to ride atop a first-stage

booster in a suborbital arc, then detach and go into orbit by firing its own restartable propulsion and maneuvering system. After completing its overflight missions, the Agena reentered the atmosphere and burned up, but not before a capsule carrying the spacecraft's camera and images returned to Earth. The air force awarded Lockheed the contract for Agena in 1956.

Agena A served several air force reconnaissance objectives: as the earliest American spy satellite, known covertly as Corona and publicly as Discoverer; and as the second stage of Atlas, in which role it boosted the missile detection and surveillance (MIDAS) early warning system and the satellite and missile observation system (SAMOS) electronic intelligence satellite. In its Corona role, Agena flew as stage two of the Thor intermediate-range (2,000-mile/3,219-kilometer) ballistic missile, developed by Douglas Aircraft for the air force and deployed mostly to the UK Air Force. Twenty Agena A stages, from January 1959 to January 1961, were launched.

Agena B had a far broader launch manifest than Agena A. Its improved Bell engine could be restarted in space and its longer body carried far more propellant, doubling the firing time to 240 seconds and leaving a surplus of fuel for orbital maneuvers. Between 1960 and 1966, Agena B flew forty-seven times on the Thor and twenty-nine on the Atlas.

NASA determined from the Corona launches that Agena B might be able to fill the void for additional lifting power for its planetary probes—especially in the 1960 to 1962 timeframe—while it waited for the more powerful Centaur to become available. Pressed also by the concern that if it delayed too long the Soviets might take the lead in lunar exploration, NASA put Agena B to work as the second stage on its Ranger moon launches. The space agency found itself relying on Agena for some time; Centaur's first launch of an actual spacecraft (Surveyor 1) did not occur until 1966. Agena D—a standardized Agena B that accepted a wide array of probes and satellites in its conical nose section—became the workhorse of the Agenas. It made 269 launches from 1962 until its decommissioning in 1987, with 125 flown on Thor, 76 on Atlas, and 68 on Titan.

NASA derived its greatest benefit from Agena D during the Gemini program. Known for this purpose as the Gemini-Agena Target Vehicle (GATV), it measured 26 feet (7.93 meters) long, nearly 5 feet (1.52 meters) in diameter, and weighed 7,180 pounds (3260 kilograms). To the basic GATV design, its project engineers added a docking collar and a status panel display. The GATV not only served as a rendezvous and docking vehicle for the Gemini missions, but it also fulfilled a third role: after docking, it retained enough fuel to act as a space tug, ferrying the Gemini capsule and crews to higher Earth orbits.

Agena Upper Stage

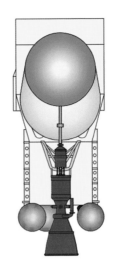

AGENA A

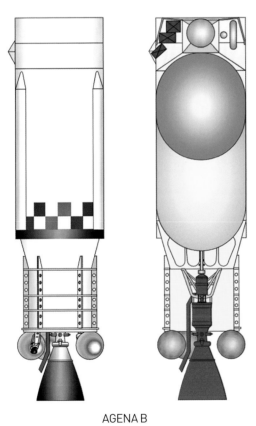

AGENA B

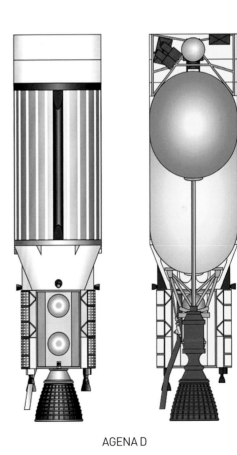

AGENA D

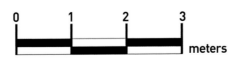

0   1   2   3

meters

The Agena target vehicle served on six missions: Gemini 6 in October 1965, Gemini 8 in March 1966, Gemini 9 in May 1966, Gemini 10 in July 1966, Gemini 11 in September 1966, and Gemini 12 in November 1966. Despite its many contributions to American reconnaissance, as well as to the early US space program as an upper-stage rocket, during its more than twenty-five-year lifespan Agena gained its greatest prominence as the GATV, the indispensable training vehicle for the Gemini astronauts as they practiced for Project Apollo.

Centaur

Taking its name from the mythological Greek figure with the trunk, head, and arms of a human being and the body of a horse, Centaur expressed a duality: the human mind coupled with equine power.

The USAF originated both Atlas and Centaur. In fact, while Convair fabricated Atlas for the USAF, it submitted a proposal in 1957 for Centaur, a powerful second-stage booster capable of lifting exceptionally heavy payloads into space. Recognizing its utility, the air force pursued the project, but in 1959 ceded it to the newly formed NASA.

NASA needed Centaur to realize its plans to explore the moon and the other planets, but its early development gave scant encouragement as delays and test stand explosions weighed down the program. When a Centaur blew up after launch in May 1962, the space agency decided to take action; it authorized the transfer of the project from headquarters to the NASA Lewis (after 1999, Glenn) Research Center field office in Cleveland, Ohio. Lewis specialized in propulsion.

Centaur became one of NASA Lewis's most significant achievements. Its engineers had been experimenting with high-energy liquid propellants since the late 1940s. For Centaur specifically, the Lewis team formulated a liquid hydrogen/liquid oxygen mixture. Through ground testing at Lewis' Plumbrook Station at Sandusky, Ohio, they improved Atlas' first stage as well.

With these innovations from Lewis in hand, Convair got to work fabricating Centaur with the same revolutionary technique that it applied to Atlas: making Centaur's tankage, rather than its bracing, the bulwark of its structural integrity. Each Centaur consisted of two thin-walled stainless-steel tanks: one for liquid hydrogen,

Centaur Upper Stage

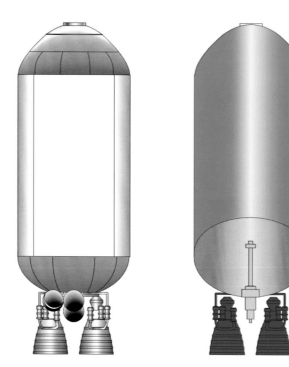

CENTAUR D

requiring sustained temperatures of -420 degrees Fahrenheit; and another for liquid oxygen, requiring -297 degrees F. Combining these propellants with two Pratt and Whitney RL-10 main rocket engines yielded 33,000 pounds total thrust.

The flight of Atlas-Centaur on November 27, 1963, marked the world's first successful use of the high-energy cocktail of liquid hydrogen and liquid oxygen, which represented a clean break from the kerosene-based hydrocarbon fuels in service on all other rockets. Centaur measured 30 feet (9.1 meters) long and 10 feet (3 meters) in diameter, with a weight of 35,000 pounds (15,876 kilograms) fueled.

In the end, after eight experimental flights—one of which ended with an explosion on the launch pad—on May 30, 1966, Atlas-Centaur served its intended purpose when it lifted Surveyor 1 out of Earth's orbit, on its way to making the first soft landing on the moon. During the 1970s, NASA engineers mated Centaur to the Titan III missile, the first flight of which failed in February 1974. But Titan-Centaur proved its mettle in the launches of the historic Viking 1 and 2 Mars missions,

respectively, in August and September 1975. Until the space shuttle era, the bulk of NASA's big planetary spacecraft relied on Centaur, including the Surveyor, Mariner, Pioneer, Viking, and Voyager probes, which visited the moon, Mercury, Venus, Mars, Jupiter, Saturn, Uranus, and Neptune. Centaur also boosted many prominent European Space Agency missions, such as the Solar Heliospheric Observatory in 1995 and Cassini-Huygens in 1997. In order to serve all of these clients, Centaur underwent improvements and modifications through the years, continuing into the twenty-first century. In all, eight different launch vehicles—Atlas-Centaur, Atlas G, Atlas I, II, III, and IV, as well as Titan III and IV—drew upon the extraordinary boosting power of Centaur.

Gemini-Titan II

During its forty-six-year lifespan, the Titan missile not only served as the Gemini launch vehicle, but also as a mainstay of the air force's orbital activities and as one of its mainline nuclear deterrents. Yet it had a surprisingly inauspicious beginning. Air

**McDonnell Gemini SC 8-12
Agena-D Docking Configuration**

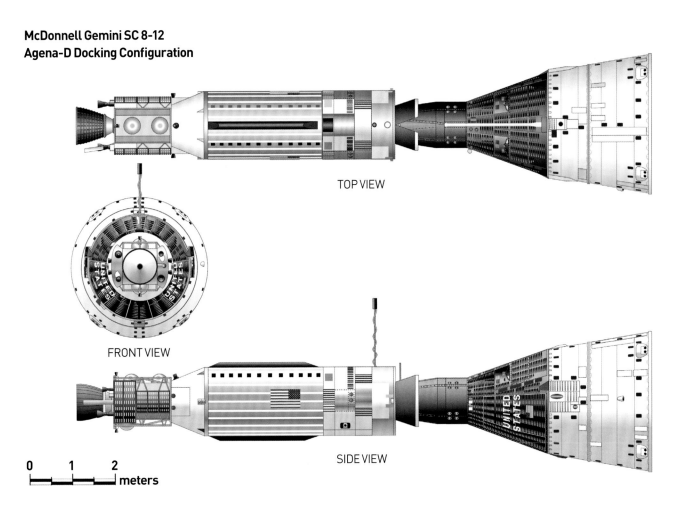

TOP VIEW

FRONT VIEW

SIDE VIEW

0   1   2   meters

Gemini-Titan II

0 1 2 3 4 5
meters

Force Gen. Bernard A. Schriever, who oversaw the testing and deployment of the Atlas missile in his role as director of the Air Force Western Development Division, championed Titan for two reasons: as an understudy for Atlas, and as a competitive goad to Atlas's builder, Convair. The Martin Company won a contract in 1955 to fabricate Titan 1, and the air force selected Aerojet to manufacture the engines. Titan 1 became operational in 1959.

But the air force recognized several weaknesses in Titan I, most notably the fifteen-minute wait to lift it from its subground silo, load its unstable propellant (liquid oxygen), and launch it. Martin proposed solutions to these and other problems with the unveiling of Titan II, fueled by easier-to-handle storable propellants and fired from inside the silo. By 1960, Convair released Titan II's characteristics: 103 feet (32 meters) long (versus 97.4 feet—30 meters—for Titan I), 10 feet (3 meters) in diameter, with 430,000 pounds of stage 1 thrust (compared to 300,000 in Titan I).

Transforming this essentially new missile into a launch vehicle for Gemini gave NASA, Martin, and Aerojet engineers plenty to handle. To begin with, the overall pattern of early Titan II launches left something to be desired. Of thirty-three research flights between March 1962 and April 1963, about nine resulted in failures or only partial successes. From NASA's viewpoint (if not the air force's), a serious incident happened on the first Titan II launch on March 16, 1962. Ninety seconds after liftoff, longitudinal oscillations occurred in the first-stage combustion chambers. The missile still traveled the required 5,000 miles and struck the target area, but this so-called "pogo" effect could only be tolerated on ballistic missiles, not on the Gemini-Titan II launch vehicle destined to carry living passengers. After several proposed solutions, project engineers agreed on measures that brought the problem (a partial vacuum in the fuel lines during pumping) under control: increasing fuel tank pressure, replacing steel feed lines with aluminum oxidizer ones, and adding a fuel surge chamber to the feed lines.

Getting Titan II cleared for human spaceflight, however, required much more than resolving one missile anomaly. Luckily, a system of control over parts and critical components—assuring that these individual pieces met the highest standards of uniformity and quality—had been established during Mercury. In fact, the procedures developed for Mercury did not just get passed on to Gemini in manuals; many of the same people who implemented the guidelines initially did so a second time for Gemini.

One of the biggest contributors to Gemini-Titan II's success involved the malfunction detection system (MDS). Unlike the one pioneered in Mercury that responded to anomalies with automatic aborts, the MDS in Gemini alerted the astronauts

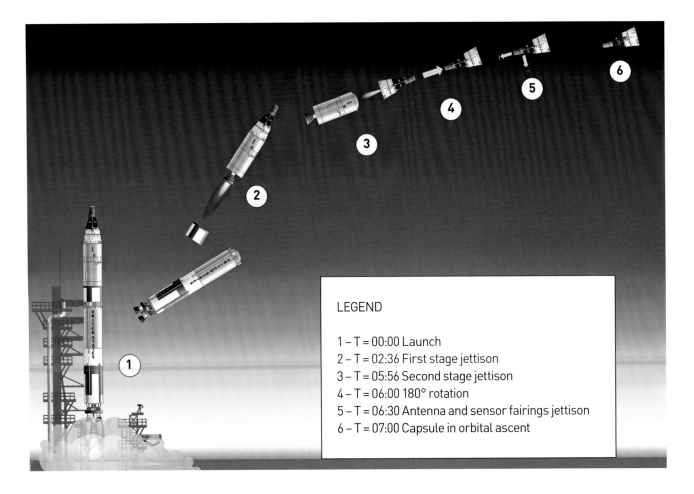

LEGEND

1 – T = 00:00 Launch
2 – T = 02:36 First stage jettison
3 – T = 05:56 Second stage jettison
4 – T = 06:00 180° rotation
5 – T = 06:30 Antenna and sensor fairings jettison
6 – T = 07:00 Capsule in orbital ascent

to problems and gave them the option of resolving them. MDS monitored such activities as the separation of the stages, changes in voltage in the electrical system, turning rates, and pressures in the thrust and the propellant tanks. Titan II also incorporated a redundant flight control system, a radio-controlled second-stage guidance system operated by ground computers, redundant electrical systems, and a forward skirt assembly that bound the spacecraft to the launch vehicle.

These modifications to Titan bore fruit. The final thirteen of the thirty-three research flights flew successfully, as did the un-crewed Gemini 1 (April 1964) and Gemini 2 (January 1965), clearing the ground for the astronauts on the Gemini 3 to 12 missions.

Saturn IB

Like so much of the military hardware adopted for the early US space program (such as Redstone, Atlas, Titan, Centaur, and Agena), the Saturn family of launch vehicles originated with the Department of Defense. In 1957, defense department officials approached the Army Ballistic Missile Agency at

Huntsville, Alabama—where Wernher von Braun and his team of German rocketeer expatriates had been located since 1950—with a request for a large booster capable of launching large space probes, reconaissance, and weather satellites. The following year the defense department's newly formed Advanced Research Projects Agency (ARPA) assumed control of the project, with the goal of developing a rocket capable of producing a staggering 1.5 million pounds of thrust (680,389 kilograms of thrust) in its first stage.

From that point onward, the Huntsville team, acting under the guidance of ARPA, became the lead agency for Saturn—which it named for the Roman God of the harvest. Meanwhile, in October 1959, the army received orders from the White House to transfer ABMA to the fledgling NASA; it reopened in July 1960 as the George C. Marshall Space Flight Center.

In a practical sense, the Saturn program at Marshall consisted of two distinct projects: the Saturn 1 and 1B, which had their roots in the pre-Apollo space program and served as moon program testbeds; and the Saturn V, which borrowed heavily from

Saturn 1B–Apollo

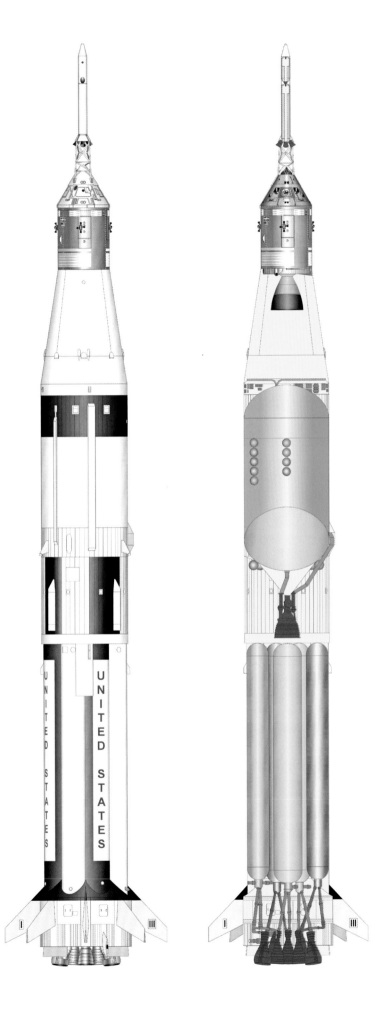

0 1 2 3 4 5 meters

Saturn V – Apollo

0 1 2 3 4 5 meters

the Saturn I and IB, but still represented something radical and new in its scale and capacities.

When ARPA became involved with Saturn, it imposed tight budgets, and the Huntsville team adapted in part by choosing proven, off-the-shelf engines for the project. Accordingly, the design of stage 1 (fabricated by Chrysler) depended on eight Rocketdyne H-1 engines, an improved version of those used in the Thor-Jupiter missiles. Its second stage, built by Douglas Aircraft, patterned itself after Centaur, with liquid nitrogen/liquid oxygen propellants and Pratt and Whitney RL10 engines (Centaur flew on two RL10s; Saturn I on six). Saturn I made its first flight in October 1961 and its tenth in July 1965, all virtually without incident. In the last four of these, it served not just as a testbed, but as a launcher. The seventh lifted a 39,000-pound (17,690-kilogram) prototype Apollo spacecraft into orbit; launches eight to ten lifted three Pegasus satellites, designed to determine the extent of meteorite strikes on later Apollo missions. In stature, Saturn I stood about midway between early missiles such as Atlas and the mighty Saturn V: 191.5 feet (58.3 meters) tall (with payload), with a maximum diameter of 21.5 feet (6.5 meters) in stage 1. As originally planned, the first stage delivered 1.5 million pounds of thrust (680,389 kilograms of thrust), with stage 2 adding an additional 90,000 pounds (40,823 kilograms).

The Saturn IB differed materially from Saturn I. Stage 1 remained all but identical, except that Pratt and Whitney further enhanced the H-I engines. But stage 2 witnessed a revolution. NASA proposed a breakthrough upper stage engine for Saturn in 1959 and awarded Rockwell the contract. The subsequent J-2—thirteen times the size of one RL10 and more powerful than all six RL10s on the Saturn I combined—represented a new generation of liquid hydrogen/liquid oxygen powerplants. The Saturn IB also had a significantly bigger profile than the Saturn I: 224 feet (68.3 meters) tall with a maximum diameter of 21.7 feet (6.6 meters). It produced 1.6 million pounds of thrust (725,748 kilometers of thrust) in stage 1 and up to 225,000 pounds (102,058 kilometers) in stage 2.

The Saturn IB's flight program allowed for thorough trials of the key Apollo systems. The first launch, in February 1966, not only marked the initial use of the J-2 engine, but put the payload—a powered Apollo spacecraft—through its paces, testing its structural integrity, communications, separation, and the command module heat shield. In contrast to this suborbital mission, in July 1966 the Saturn IB underwent its maiden orbital flight with a 58,500-pound (26,535-kilogram) mixed payload, the heaviest cargo hauled by the United States to date. A planned simulated restart of the J-2 engine occurred on this flight. Another suborbital mission followed in August, during which

the command module heat shield underwent the ultimate test: it rose to 2,700 degrees Fahrenheit on reentry, but the cabin temperature stayed cool at 70 degrees.

Then, the Saturn IB became associated with tragedy when Apollo 1—the first mission with astronauts on board, scheduled for launch on February 21, 1967—caught fire on launch pad 34 at Kennedy Space Center on January 27, killing astronauts Gus Grissom, Edward White, and Roger Chaffee. After long months of investigation, the program resumed in October 1968 with Apollo 7, flown by Walter Schirra, Walter Cunningham, and Donn Eisele. The astronauts tested the flight worthiness of the command and service module in preparation for the lunar circumnavigation scheduled for Apollo 8, including a simulated rendezvous and docking with the second stage. The Saturn IB ended its lunar career with Apollo 7, but supplied the transportation for the astronauts on Skylabs 2, 3, and 4 (launched in May, July, and November 1973), and finally, in July 1975, carried the American crew of the Apollo-Soyuz Test Project to meet their Soviet counterparts.

Saturn V

As early as 1955, the air force contracted with the Rocketdyne division of North American Aviation to answer a simple question: What maximum thrust limitations existed for liquid propellant engines? The contractor responded that a million pounds of thrust seemed possible and supported the claim by submitting a preliminary design. The air force requested that Rocketdyne conduct further research and development at its Canoga Park, California, plant and to run tests at its Edwards Air Force Base facility. With the establishment of NASA, the air force transferred the project to the space agency, which in January 1959 contracted with Rocketdyne for an even more powerful engine, capable of *1.5 million* pounds of thrust (680,389 kilograms of thrust)—even though no US rocket of the time could begin to accommodate it.

By spring 1961, Rocketdyne static-tested a prototype engine that delivered an incredible 1.64 million pounds of thrust (725,748 kilometers of thrust). Called the F-1, it represented not so much a breakthrough in propulsion technology (it used liquid oxygen and kerosene as propellants), but as a massive improvement of earlier designs. The bell-shaped powerplant measured 19.8 feet (almost 6 meters) in length and 12.3 feet (3.76 meters) in diameter. But its vast size and power inevitably brought problems: during its development, the F-1 experienced twenty failures—nine of them explosions—in its turbopump mechanism.

By the close of 1961, NASA decided how it wanted to apply the F-1 technology. It committed itself to the first stage of an immense launch vehicle, capable of 7.5 million pounds of thrust (3,401,943 kilometers of thrust), achieved by five F-1s (four outboard and gimbaled for steering; one at the center). A preliminary contract with Boeing signed in February 1962 constituted the birth certificate of the Saturn V rocket, Apollo's ride to the moon.

Saturn V's second stage proved to be more problematic than the first due to management disputes between the contractor and the von Braun team. Although it used the same new and powerful J-2 engine as the Saturn IB, controversy arose not about the J-2, but about the upper stage itself. NASA asked for a structure 74 feet (22.5 meters) long and 21.5 feet (6.4 meters) in diameter with exacting specifications—in other words, Marshall requested not only an imposingly big structure, but one that required the tolerances of a Swiss watch on an object the size of a locomotive. North American-Rocketdyne fell behind on its deliveries and the budget grew, causing disagreements to break out between the Marshall engineers and the contractor. The persuasive and well-connected James Webb applied himself to the problem; he pressured the company's CEO, management changes ensued, and by late 1967, stage two's development came into focus.

Onlookers who saw the Saturn V on pad 39A at Kennedy before its first launch witnessed something that had to be seen (and heard) to be believed. A towering 363 feet (111 meters) tall and with a diameter of 33 feet (10 meters), the rocket weighed 6,400,000 pounds (2,903,000 kilograms) at liftoff and could carry a translunar payload of about 107,350 pounds (48,693 kilograms). Its stages, from one to three, respectively, developed 7,610,000, 1,150,000, and 230,000 pounds of thrust (3,451,838, 521,631, and 104,326 kilograms of thrust); in all, 8,990,000 pounds (4,077,795 kilograms), compared to 1,825,000 pounds (827,806 kilograms) for the Saturn 1B.

On its initial flight, the Saturn V performed well, except for some low level longitudinal oscillations (a "pogo" motion) emanating from the F-1 engines. However, during Apollo 6 in April 1968, just before the cutoff of the F-1s, the pogo effect became much more severe. Additionally, during this mission, one of the J-2s in stage 2 failed, and the J-2 in stage 3 did not start. Teams investigating these problems issued their findings and technical changes resulted. But surprisingly, NASA officials decided against another test flight. Instead, with the clock running out on achieving President Kennedy's 1969 target, the space agency jumped the schedule and in December 1968 put three astronauts on board Apollo 8 for a circumnavigation of the moon—a big gamble that paid off handsomely. The Saturn V performed reliably, ultimately carrying twenty-one Americans on lunar landing missions. The huge rocket flew a total of twelve times during Apollo.

After Apollo ended, the Saturn V made two more contributions: it launched Skylab 1 into orbit in May 1973 and it also served on board Skylab itself; a refashioned Saturn V second stage acted as the shell of the lab's orbital workshop.

R-7

The USSR derived its first rocket—the R-1—from the propulsion design of captured German V-2s (as well as from German scientists, documents, and equipment rounded up after the war). The R-1 flew for the first time in 1947, following which the Soviets launched the R-2 (double the range of the R-1), the R-5 (the first Soviet missile able to carry a nuclear warhead), and the R-11 (a smaller surface-to-air missile).

As a new and more powerful rocket emerged from the design bureau—which was led by the Soviet Union's legendary engineer and space pioneer, Sergei P. Korolev—his friend Mikhail Tikhonravov approached him with a novel idea. He asked Korolev to consider adapting the new missile—known as the R-7, or Semyorka—for a dual role as a spacecraft launch vehicle. As basic research on Korolev's missile progressed during the late 1940s and early 1950s, he modified the design to satisfy Tikhonravov's request, enabling it to haul a heavier payload than the warhead envisioned originally for the R-7.

He followed up in 1953 by proposing to the Communist Party Central Committee that the R-7 serve as a satellite launcher, in addition to its primary role as an ICBM. The Soviet Ministry of Defense approved the R-7 in June 1954 (after the R-7 required a hurried modification to accommodate the new thermonuclear warhead). Then, in 1955, Russia accepted the challenge of the International Geophysical Year (IGY) organizers to launch an Earth satellite in the 1957–1958 timeframe. To expedite this goal, the Soviet authorities subordinated the Tikhonravov group to Korolev's design bureau in 1956.

Korolev's collaboration with Tikhonravov bore fruit in the first few days of October, 1957, as crews at the Baikonur Cosmodrome in Kazakhstan prepared the R-7 for a revolutionary mission. The technicians handled no nuclear materials for this flight, only an unassuming, 184-pound (83.6-kilogram) polished aluminum sphere just 23 inches (58 centimeters) in diameter. The rocket itself measured 98 feet (30 meters) in height, 9.0 feet (2.99 meters) in diameter, and weighed 588,000 pounds (267,000 kilograms) at liftoff. It relied on kerosene and liquid oxygen propellants, fed into a first stage consisting of one RD-108 engine, and into four boosters, each with a single RD-107 engine. Although the Semyorka carried a light load on this occasion, it could transport up to 1,100 pounds (500 kilograms) to low Earth orbit.

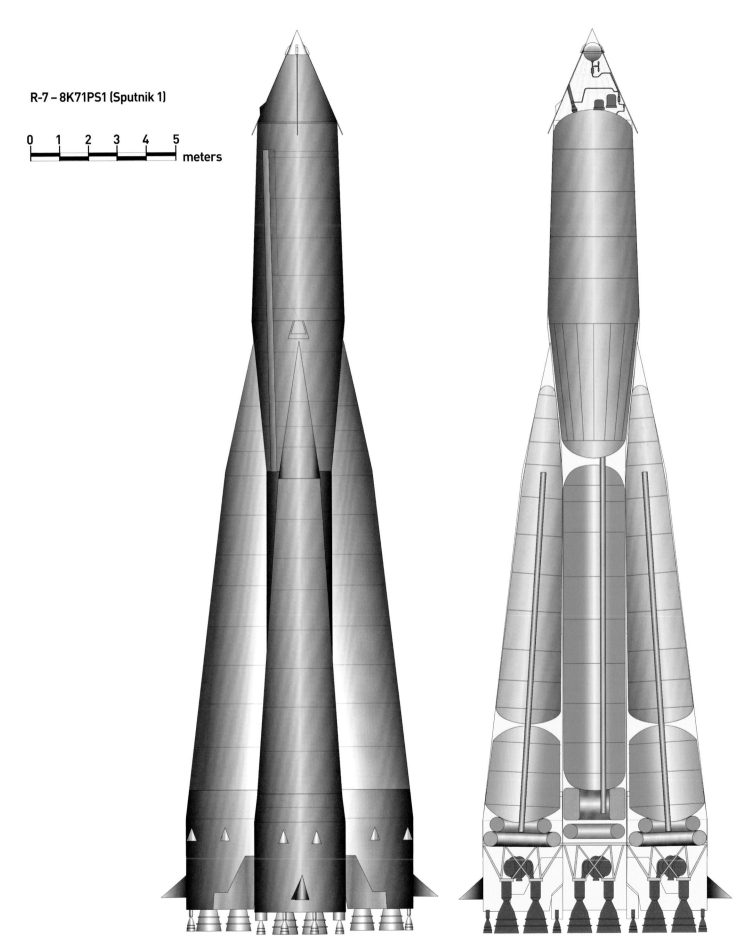

R-7 – 8K71PS1 (Sputnik 1)

0 1 2 3 4 5
meters

N1 Launcher

N1/7L   N1/L1S   N1/L3

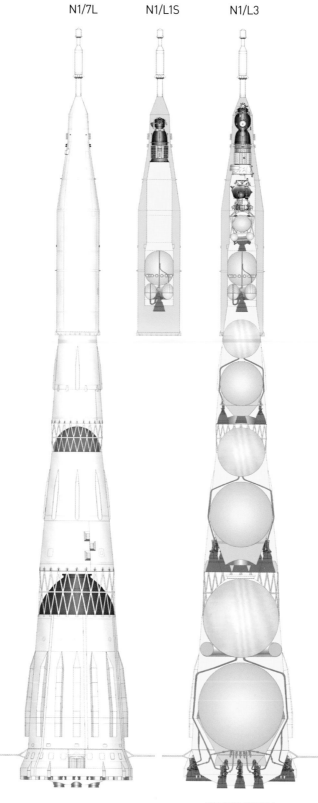

0 1 2 3 4 5 meters

SIDE SECTION

The R-7's successful launch of Sputnik 1 on October 4, 1957, touched off a chain reaction that brought profound consequences: not only did it incite near panic among American politicians and initiate a bruising competition with the United States for supremacy in space, but it also poured fuel on the Cold War by intensifying an ideological rivalry in which both sides sought to prove its political superiority to a worldwide audience.

More narrowly, as a class of launch vehicles, the R-7 proved to be exceptionally successful. They flew for more than fifty years, and the derivative Soyuz rockets continue to carry crews and supplies to the International Space Station in the twenty-first century.

N1

The USSR's biggest rocket program began with three obstacles: insufficient early funding, a late start, and the loss of the most important figure in the Soviet space program to guide it.

Like so many of the Soviet efforts, the N1 originated in Special Design Bureau 1, led by Sergei P. Korolev. Korolev and his team conceived of the N1 in 1959 as a booster for military launches and for manned flybys of Venus and Mars. In 1961, he got limited funding for two years of research. Meanwhile, well aware of US ambitions for a moon landing, Korolev persuaded Leonid Brezhnev—the general secretary of the Communist Party—to endorse a moon shot. But the final go-ahead came late, in October 1965, nearly four years after initiation of the Saturn V program. And just three months after the formal approval, Korolev died. His deputy and successor, Vasily Mishin, lacked his late bosses' connections and political acumen, and so the N1 started out as a crash program without Russia's most able and experienced leader at the controls.

On top of these formidable difficulties—added to which, Korolev had promised to build not just a new rocket, but a new capsule (the Soyuz 7K-LOK) and a lunar lander (the LK)—his engineering staff conceived of a launch vehicle of immense proportions and power never before attempted by the Soviets. The N1 measured 344 feet (105 meters) long (including payload), with a diameter of 55.8 feet (17 meters), and a mass of 6,060,000 pounds (2,750,000 kilograms) fueled. It could lift a cargo of up to 209,000 pounds (95,000 kilograms) into low Earth orbit. Its four stages contained forty-three engines in all, fed by Refined Petroleum-1 (kerosene) and liquid oxygen, with thirty NK-15s in stage one, arranged in two rings; eight NK-15Vs in stage 2, laid out in a single ring; four NK-21s in stage 3; and one NK-19 in stage 4.

In the final analysis, the N1 fell victim to flaws in its design and to a launch schedule determined less by technical readiness than by Apollo's progress. Because of the decision to mount its engines in clusters, the plumbing carrying the propellant and oxidizer proved to be overly complex and fragile. Moreover, no N1 underwent a test flight, nor did the thirty engines in stage 1 undergo static firings as a unit.

The results turned out to be catastrophic. All four N1 flights were fortunately unmanned as during the first, on February 21, 1969, pogo oscillations at six seconds after launch ripped some components off their moorings and caused a propellant leak. The first stage shut down at 68 seconds and the N1 crashed to the ground at 183 seconds. Flight number two, on July 3, 1969, appeared to be normal for the first ten seconds, when mission control noticed that pressure fell to zero on engines 1 to 12. The liquid oxygen turbopump in engine 8 exploded, starting a fire. Then, at 10.5 seconds and at an altitude of about 328 feet (100 meters), the rocket seemed to freeze, tilt to one side, and fall back on the launch pad, at which point a massive red and black cloud erupted. One military observer at the site said, "Today . . . I saw without exaggeration the end of the world, and not in a nightmare, but while fully awake and standing right next to it." He experienced one of the greatest nonnuclear explosions in human history.

After a break of almost two years (the schedule could now relax; Apollo 11 landed on July 20, 1969) came the launch of N1 number three, on June 26, 1971. Soon after liftoff, the rocket began to roll and at forty-eight seconds, it disintegrated. And in the fourth flight, on November 23, 1972, all seemed well until ninety seconds elapsed, at which point a routine shutdown of the six center engines in stage 1 caused excessive dynamic loads. That force burst the propellant lines, causing a fire. The first stage broke up at 107 seconds.

The N1 program came to an end in May 1974. It represented a bitter chapter in the history of a Soviet space agency that only a few years earlier had been lionized for one stunning space first after another. At least partial redemption lay ahead in the Salyut and Mir space stations, and eventually a new era of spaceflight dawned when the US and Soviet space agencies merged their talents during the 1990s to conceive of the International Space Station.

UR-500 Proton

If Soyuz represented perhaps the most durable and long-serving series of spacecraft that transported people into space, the Proton rocket should rightfully be characterized as the Soyuz of the launch pad. Initiated more than fifty years ago, these two veterans of the early Space Age continue their work almost two decades into the twentieth-first century.

**Proton Launcher (8K82K) Zond 4-8
(Soyuz 7K-L1) Launch Configuration**

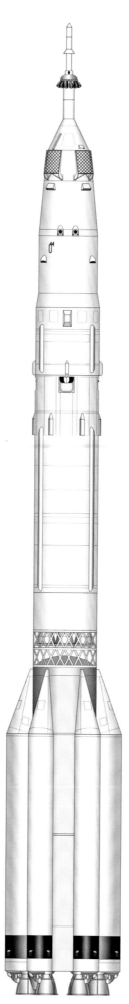

0 1 2 3 4 5
meters

Ironically, one of the world's most effective rockets began as a counterweight to the Soviet space program. During the early 1960s, many military leaders in the USSR felt that Premier Nikita Khrushchev diverted too much of the defense budget into space activities. Accordingly, a government decree issued in May 1961 reduced the Soviet space effort. Adapting to the mood of the time, Vladimir Chelomei, the chief designer of Union Design Bureau 52 (OKB-52), proposed a massive, 500-ton ballistic missile capable of delivering a 100-megaton super bomb to a target.

Khrushchev approved the project in April 1962, but with his fall from power in October 1964, Chelomei lost his main protector and the big rocket faced extinction. The axe failed to drop because Mstislav Keldysh, the head of the Soviet Academy of Sciences, chaired a commission in August 1965 that decided to save it as a booster for Russia's proposed LK-1 circumlunar capsule.

Almost accidentally, then, the Proton became Russia's biggest effective launch vehicle and the country's response to the America's Saturn V. It weighed 1,312,830 pounds (595,490 kilograms) and could lift a payload of 18,500 pounds (8,400 kilograms). The UR-500 measured nearly 152 feet (46.28 meters) in length with a diameter of 13.61 feet (4.15 meters). Its first stage clustered six 11D43 engines (designed originally for the N1 rocket); the second stage had four modified engines from the first stage of OKB-52's UR-200 ballistic missile. The original UR-500 flew just four times: in July and November 1965 and in March and July 1966. Each of the flights attempted to launch an x-ray satellite, weighing 18,200 pounds (8,300 kilograms) apiece, at the upper end of the Proton's lifting capacity. Despite an oxidizer leak, the first one succeeded. The second and fourth also launched their payloads, but on the third try (on March 24, 1966), a second stage malfunction ended the mission.

The most historic part of the Proton story unfolded after the initial UR-500. From that point until the present, it has been adapted again and again for a wide variety of missions. It has sent planetary probes to Venus, Mars, and the moon, and placed satellites into geostationary orbit. With the addition of a third stage (as Proton-K and -M), it launched every Soviet space station, in addition to heavy-transport spacecraft. During the 1990s, Protons took on the role of commercial booster, sending its first one into space in 1996. A fourth Proton stage—known as the Briz-M, initiated in 2001—enabled the Soviet space program to lift cargoes beyond low Earth orbit.

Despite setbacks and occasional failures, the UR-500 and its derivatives earned a solid record. Between 1965 and 2016, they completed 365 of 412 launches for an 88.6 percent success rate.

ROBOTICS
Explorer 1

The first American satellite—the spacecraft that brought the United States into the Space Age—originated as a weird hybrid: as a scientific endeavor and as a byproduct of Cold War intrigue.

It began when the International Council of Scientific Unions declared in 1952 an International Polar Year (IPY), to occur from 1957 to 1958, with the intention of mapping the most remote parts of the Earth. But soon its organizers renamed it the International Geophysical Year (IGY) after inviting interested countries to commit themselves to pursuing orbital spacecraft for Earth observation. The US and Soviet governments, both developing missiles as launch vehicles for nuclear weapons, took the challenge seriously.

After a keen competition and detailed satellite proposals from the army, air force, and navy, the Department of Defense chose the Naval Research Laboratory's Vanguard as the American response to the IGY call. It won because, as a nonmilitary research project, it did not compete for resources with the defense department's high priority ballistic missile program. Additionally, the navy enjoyed great success with Vanguard's predecessor, the Viking sounding rocket. But almost from its inception in 1955, Vanguard encountered problems, culminating in a disastrous launch pad failure on December 6, 1957. This well-publicized event—as well as the stunning success of Sputnik 1 on October 4, 1957, followed by Sputnik 2 in November—persuaded the DoD to continue with Vanguard, but to reawaken and push forward the army's earlier satellite plan.

The army jumped at the chance. It had awaited this moment since 1954 and 1955 when, under Project Orbiter, the Army Ballistic Missile Agency in Huntsville, Alabama, and the Jet Propulsion Laboratory (then an army contractor) in Pasadena, California, joined forces to propose a satellite and launch vehicle. When Vanguard got the assignment, Project Orbiter ended. But when this new opportunity arose, the Huntsville engineers began to adapt the well-tested Redstone rocket (a direct descendant of the V-2 from World War II) as a first stage booster. JPL, meanwhile, got to work on the three upper stages, the communications systems, and the satellite itself. These institutions on opposite sides of the country shifted into high gear in late 1957 when the Eisenhower administration gave them just eighty days to orbit the first American satellite.

In its final form, the satellite—weighing just over 30 pounds (14 kilograms)—consisted of a steel cylinder 80 inches (203 centimeters) long and 6 inches (15 centimeters) in diameter.

Explorer 1 (1958 Alpha 1) Satellite

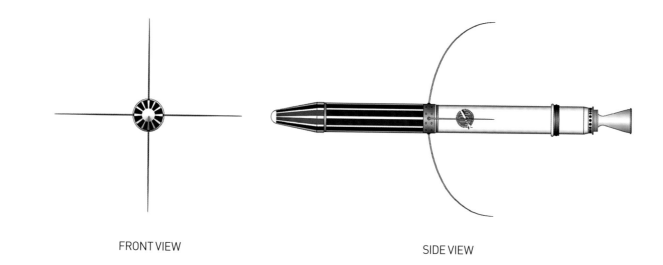

FRONT VIEW   SIDE VIEW

0   5

meters

Vanguard 1 (1958 Beta) Satellite

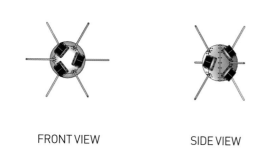

FRONT VIEW   SIDE VIEW

0   5

meters

Its battery-powered transmitters sent data back to Earth and provided a tracking signal.

On board the spacecraft, Professor James Van Allen's experimental cosmic ray detector made one of the most important discoveries of IGY. Once in orbit, his instruments found the anticipated cosmic rays, but at a much lower concentration than expected. Van Allen guessed that his equipment had been overwhelmed by charged particles embedded in the Earth's magnetic field. The full extent of the phenomenon, subsequently proved by Explorer III, became known as the Van Allen radiation belts.

Vanguard 1

Unlike the development of Explorer 1—in which program managers struggled to cope with an eighty-day deadline to launch their satellite—those involved in Project Vanguard enjoyed both the luxury and the curse of time.

Vanguard (a name applied both to the satellite and the rocket on which it flew) originated with an announcement made by the International Council of Scientific Unions in 1952, designating 1957 and 1958 as the International Geophysical Year (IGY). Eager to encourage the study of the most remote parts of the Earth's surface, the IGY's organizers issued a call: to explore the world using artificial satellites. The Eisenhower administration directed the Department of Defense to ask the three military services for proposals to meet the IGY's challenge. In 1955 the defense department selected the navy's Vanguard over the army and air force candidates. The decision reflected Vanguard's focus

Vanguard TV-4 (Vanguard 1)

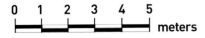

0 1 2 3 4 5 meters

on research, not military utility, and reflected confidence in the navy's successful predecessor to Vanguard, the Viking sounding rocket. But additionally, President Eisenhower intervened, arguing that the armed forces' ballistic missile programs must not become entangled in IGY activities. He feared that this civilian pursuit would divert time, talent, and resources from the earliest possible deployment of ICBMs. That decision proved to be fateful.

The Naval Research Laboratory (NRL) in Washington, DC, managed the Vanguard program, and it soon found itself inundated with potential science experiments for its satellite. Among the most prominent, astrophysicist James Van Allen of the University of Iowa proposed a space-borne cosmic ray detector to measure the extent of charged particles surrounding the Earth. But his project required a cylindrical-shaped casing, at odds with the spherical design preferred by the NRL. Also, Van Allen's equipment weighed too much; the Vanguard rocket's limited thrust imposed severe restrictions on payload, allowing no more than 22 pounds (10 kilograms) for the entire satellite and about 2 pounds (one kilogram) for the science instruments themselves.

But Vanguard's leader, Dr. John Hagen, did not lack other science options. One research team proposed a package that measured meteoric dust erosion; another sought to determine the variation of solar radiation intensity with each revolution of the satellite. By late 1955, five proposals lay on Hagen's desk, and after a Vanguard symposium in January 1956, the number rose to fifteen. But an even bigger issue than payload pressed on Hagen and his group. After some debate, they decided to take a gamble on their satellite's electrical source; NRL became the first to adopt solar-powered batteries for a spacecraft. They recently had been developed at Bell Labs and tested at the Army Signal Corps Engineering Laboratory.

The wager paid off, but in general the path of Vanguard did not run smoothly. Just as the rocket lifted off on December 6, 1957, it became enveloped in a massive fireball, which attracted international media attention. After this, the department of defense directed the army to revive its earlier satellite proposal and to proceed with the development of an orbital spacecraft. Ultimately, the Army's Explorer 1 became the first US satellite in space, launched on January 31, 1958. Vanguard came second when it reached orbit on March 17, 1958—St. Patrick's Day. Fashioned out of aluminum with six prominent antennae, the tiny Vanguard sphere measured only 6.5 inches (16.5 centimeters) in diameter and weighed a mere 3.2 pounds (1.46 kilograms).

In the end, Hagen and his colleagues at NRL chose an uncomplicated science experiment for Vanguard 1: a radio phase-comparison angle tracking system that used two transmitters: one for telemetry powered by mercury batteries, and the other a Minitrack beacon transmitter powered by six solar cells mounted to the exterior of the spacecraft. The first solar cells ever used in space, they continued to produce electricity until May 1964. By compiling tracking data, the experiment revealed something suspected by physicists, but unproven until now: that rather than being round with flattened poles, the Earth appeared to be pear-shaped, with the stem at the North Pole. Additionally, Vanguard revealed to researchers that the gravitational pull of the sun and moon, as well as the radiation pressure of solar light, affected the orbit of satellites around the Earth.

Despite its hard road to success, nearly sixty years after its launch, Vanguard 1 continues to revolve around the Earth, the oldest man-made object in space.

Lunar Orbiter 1

Once scientists discovered that the composition of the moon's surface could support landings, they turned to the next logical step for human exploration: finding potential touchdown points.

To achieve this goal, NASA selected its Langley Flight Research Center in Hampton, Virginia, to manage a contract with the Boeing Company for a suite of identical satellites to map the moon's surface. In sharp contrast to Ranger's six aborted missions, all five of these vehicles, known as the lunar orbiters, succeeded at least in part in photographing most of the moon's surface, with a resolution of at least 200 feet (60 meters).

Lunar Orbiter 1 lifted off on an Atlas-Agena rocket on August 10, 1966, and took pictures from August 18 to 29. Its one camera viewed the targeted territory through two separate lens systems: an 80-millimeter for medium resolution, and a 610-millimeter for high. The spacecraft had a somewhat squat, unassuming profile, like a short cone, about 5.4 feet (1.65 meters) tall and 4.9 feet (1.5 meters) across at the base. It weighed about 852 pounds (385.6 kilograms).

While not perfect, Lunar Orbiter 1 achieved its basic purpose. It took 187 medium- and 42 high-resolution images over 1.9 million square miles (5 million square kilometers) of lunar landscape, accomplishing roughly 75 percent of the intended mission. Unfortunately, many of the initial high-resolution pictures showed extensive smearing, but the rest of the images pinpointed nine prime and seven subsidiary touchdown sites for Apollo, and one for the prospective soft lunar landers, known as Surveyors. Lunar Orbiter 1 also determined that the Apollo hardware could, in the short term, counteract the radiation on the surface of the moon well enough to protect the astronauts.

Like all the spacecraft in this series, Lunar Orbiter 1 ended its mission by impacting the moon, an intentional choice designed to avoid potential collisions with the later Apollo flights. But before it

Lunar Orbiter 1

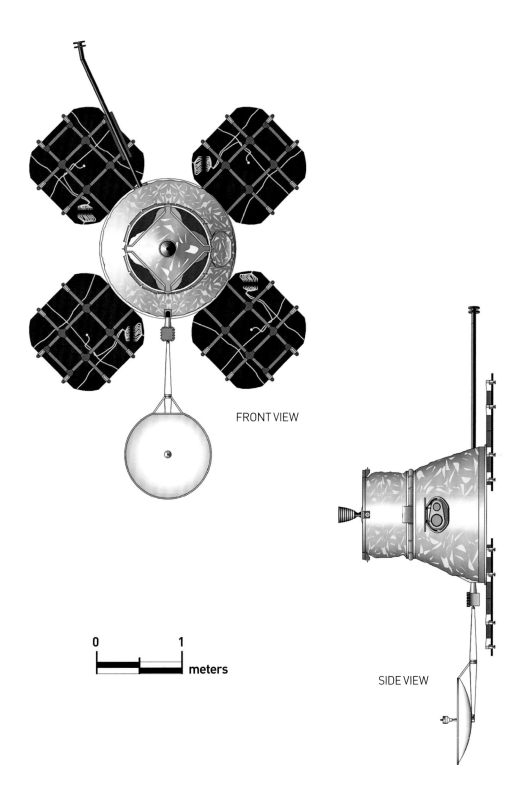

FRONT VIEW

0   1
meters

SIDE VIEW

Ranger 7 (Batch 3) Probe

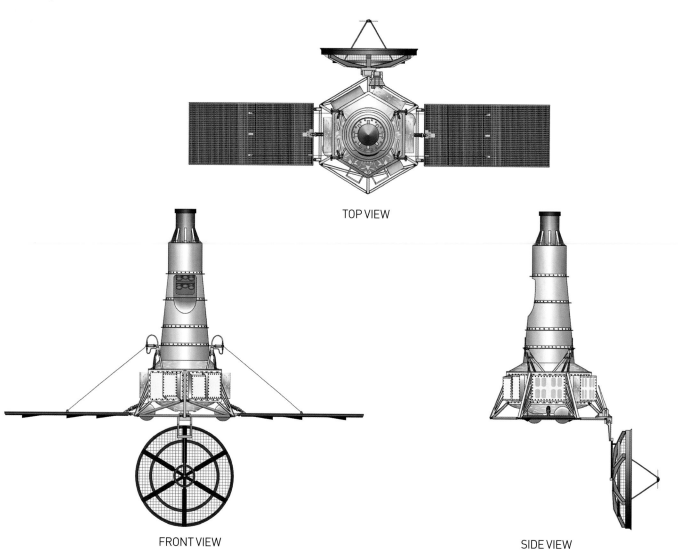

TOP VIEW

FRONT VIEW

SIDE VIEW

0   1

meters

did, controllers turned its camera homeward and it took the first two American photographs of Earth from the moon.

The voyages that succeeded the Lunar Orbiter 1 flew from November 1966 to August 1967. Lunar Orbiter 2, launched on November 6, 1966, and Lunar Orbiter 3, which lifted off on February 6, 1967, both captured images of relatively smooth terrain with good prospects for the Surveyor, as well as for the Apollo landings. Between them they returned 1,433 exceptionally sharp images, with resolution down to 1 meter. Satisfied that Lunar Orbiters 1 to 3 provided sufficient data to map locations for later touchdowns, NASA mission planners programmed Lunar Orbiter 4 to pursue a different objective: to make a systematic survey of the moon's geologic features for future scientific research. It swept over the lunar surface in May 1967, capturing more than five hundred images that constituted 99 percent of the moon's near side and 95 percent of its far side. To end the project, in August 1967 Lunar Orbiter 5 completed the final Surveyor and Apollo site selection photography, and it also took pictures of the remaining 5 percent on the moon's far hemisphere.

Ranger 7

Before the United States could fulfill the promise of President John F. Kennedy to land astronauts on the moon, NASA mission planners faced a welter of technological problems, any one of which might have ended the mission. One of the most underappreciated did not involve technology, but geology. No one really knew for certain the composition of the moon's surface; if powdery, would great amounts of dust be disturbed in a landing? Would it impair visibility, hovering like a cloud around the spacecraft? And even if this phenomenon failed to occur, did the moon's crust have enough density to support the weight of whatever touched down on it, or did it have a consistency so soft or spongy that it would topple any visiting spacecraft in a heap? Scientists also wanted to know whether the distance from Earth to the moon (about 239,000 miles or roughly 384,400 kilometers) might inhibit communications with the astronauts, and whether radiation posed a risk to their health.

To find the answers, NASA designed three types of lunar spacecraft: a crash-lander (Ranger), an orbiting scanner (lunar orbiter), and one that made a controlled descent (Surveyor).

The Jet Propulsion Laboratory (JPL) in Pasadena, California, conceived and fabricated the Ranger. Yet, despite JPL's solid reputation—it pioneered American rocketry with the Corporal and Sergeant and later created the first US satellite with Explorer 1 and its upper stages—the Ranger team experienced almost two and a half years of uninterrupted misfortune. The first two in the series, launched in August and November 1961, could not escape

Earth's orbit. Ranger 3 flew the following January, but mission control lost contact with it on the way to the moon, and Ranger 4's solar panels failed to open before it impacted the moon in April 1962. In October 1962, Ranger 5 repeated the outcome of Ranger 3; and, after a fifteen-month pause in activity, Ranger 6's cameras took not a single photograph before the spacecraft plowed into the lunar surface at the end of January 1964.

But finally, almost three years after Ranger began, it started to pay dividends. The ultimate, block-III version (Rangers 6 to 9) weighed about 807 pounds (366 kilograms), significantly heavier and more complex than block II (Rangers 3 to 5) at 728 pounds (330 kilograms), or block I (Rangers 1 and 2) at 670 pounds (304 kilograms). Launched from an Atlas Agena booster on July 28, 1964, Ranger 7 resembled a tall cone resting on a low base. Its hexagonal aluminum platform, just about 5 feet (1.5 meters) across, held the propulsion and power units; six cameras occupied the tower. It approached the moon precisely on target, and fifteen minutes before hitting its mark, its cameras captured the first of a total of 4,316 photographs of the touchdown zone, in Mare Cognitum (The Sea That Has Become Known). The evidence regarding the moon's surface continued to mount early the next year when, on February 20, 1965, Ranger 8 captured 7,137 pictures of NASA's prime landing spot for Apollo, the Sea of Tranquility. To end the program, Ranger 9 took five thousand pictures as it descended on March 24, 1965, into the crater Alphonsus, a site of geological interest, but not relevant to Apollo.

Now possessing almost 16,500 detailed images, each one thousands of times higher resolution than any telescope could capture from Earth, NASA's geologists and engineers knew by the mid-1960s that Apollo could land safely on the lunar surface.

Surveyor 1

In its attempts to study the moon prior to the Apollo landings, NASA produced three types of robotic spacecraft with radically different missions: Surveyor was the one built for controlled touchdowns. Not unexpectedly in the period just after the Space Age dawned, the accident rates of these three programs varied widely. Ranger tried for almost three years with six consecutive failures before achieving its objective; the five lunar orbiters, on the other hand, all flew successfully (but not flawlessly). Surveyor achieved a midpoint between its two predecessors: seven attempts, two losses.

In addition to the limited experience of spacecraft designers at this early stage, these differences may in part be related to the complexity of the missions. Clearly, an intact landing posed the biggest technical challenge, that carried the greatest consequences for future space travel. If human beings hoped to explore the moon and the planets, their spacecraft required safe and reliable ways to descend onto these new worlds.

NASA's candidate to develop this technology resembled a low-slung piece of photographic equipment. Surveyor measured almost 10 feet (3 meters) tall and sat on a broad tripod base fashioned from aluminum tubing with a spread of about 14 feet (4.3 meters). A mast jutted out from the top of the structure, holding two small solar panels. Its legs—equipped with shock absorbers—folded prior to liftoff into the nose shroud of its Atlas-Centaur launch vehicle. It weighed 2,194 pounds (995.2 kilograms) at liftoff.

Much rested on the success of Surveyor. Above all, it needed to scout potential landing sites for Apollo. But in the service of this mission, its designers also expected it to perform accurate midcourse and terminal maneuvers; validate the technology required for soft touchdowns; prove the value of the communication system and the Deep Space Network; verify its booster's ability to put Surveyor on a trajectory to intercept the moon; and contribute to the scientific knowledge of the moon.

Surveyor succeeded on its first try. Following Surveyor 1's launch on May 30, 1966, it achieved the world's first *controlled* landing, settling down on the Ocean of Storms. Over the following six weeks (June 2 to July 13), it transmitted 11,240 high-resolution photographs depicting a range of images, from the horizon to closeups of its own mirrors.

Surveyor 2 fared less well. It crashed on September 23, 1966, southeast of the Crater Copernicus when one of its vernier engines failed to fire, causing the spacecraft to tumble. Surveyor 3 did better, but with its own problems. It also descended to the Ocean of Storms, but because its engines did not shut off as planned, it bounced three times. It moved almost 102 feet (31 meters) off target—before stopping. It still captured 6,315 images and became the first probe ever to excavate a planetary body, scooping out bits of the moon's surface over a total of eighteen hours and showing the contents to the camera. It remained active from April 20 to May 4, 1967. Surveyor 4, launched on July 14, 1967, broke up on the moon's Central Bay after NASA lost radio contact.

The final three spacecraft in this series performed satisfactorily, although Surveyor 5 encountered a major issue with a helium regulator leak, which required mission control to plot out an unorthodox descent profile. This improvisation enabled it to come to rest on the Sea of Tranquility, where it took more than eighteen thousand images and made the first on-site chemical analysis of lunar soil. Surveyor 6 thrilled mission control with its almost error-free performance. Landing in the Central Bay, between November 10 and 26, 1967, it captured almost thirty thousand images. Perhaps equally important, as Surveyor 6

stood on the moon, its handlers on Earth fired its vernier engines for two and a half seconds, making it jump 8.2 feet (2.5 meters) from its starting position—a useful maneuver to have for later Apollo missions. Finally, Surveyor 7, launched on January 7, 1968, reached the moon on a science mission. It dropped onto the lunar landscape near Tycho Crater, where chemical analysis (and twenty-one thousand photos) suggested that debris once flowed there in a molten state. It also detected a 1-watt laser beam flashed from Earth, designed to test a new form of communication in space.

Despite occasional mishaps, Surveyor represented a decisive step forward in spacecraft capability and clinched the argument that the moon's geology offered no inherent obstacles to landing manned spacecraft on its surface.

Mariner 10

At roughly the same time that the Pioneer probes concentrated on long-distance interplanetary travel and on solar exploration, the Mariner spacecraft investigated the Earth's neighboring planets—Venus, Mercury, and Mars. The Mariners also followed a pattern established in the Explorer and Ranger programs and continued by many other American robotic spacecraft later on: rather than contract with an aerospace manufacturer, the Jet Propulsion Laboratory (JPL) conceived and fabricated the Mariners on campus and also provided mission control.

The Mariners started at a low point in JPL history. The project evolved from the predecessor Ranger spacecraft, whose record between August 1961 to January 1964 did nothing to inspire confidence. The first six Rangers—all intended to crash land on the moon's surface—failed to satisfy their objectives. As NASA headquarters and the political forces in Washington, DC, became restless for a success, JPL tried a new approach: the team lightly reconfigured a Ranger spacecraft, called it Mariner 1, and in July 1962 launched it on a flyby of Venus—only to be foiled again as it veered away from its chosen trajectory. Then JPL took an enormous gamble to save its sinking reputation—in just a thirty-six-day turnaround, its engineers took a Ranger spacecraft out of storage, outfitted it with the same suite of instruments as Mariner 1, and sent it to Venus in August 1962. After a jittery launch in which its Atlas-Agena rocket went into a galloping roll—followed by a string of hair-raising incidents, any one of which could have aborted the mission—Mariner 2 became the champion that JPL needed. It scanned Venus during a forty-two-minute flyby in December 1962 and found a thick cloud cover and exceptionally high surface temperatures (797 degrees F, 425 degrees C).

Even before Mariner 2 became the first successful mission by any country to another planet, JPL engineers began planning

Surveyor 1 (Surveyor-A) Probe
Prelanding Configuration

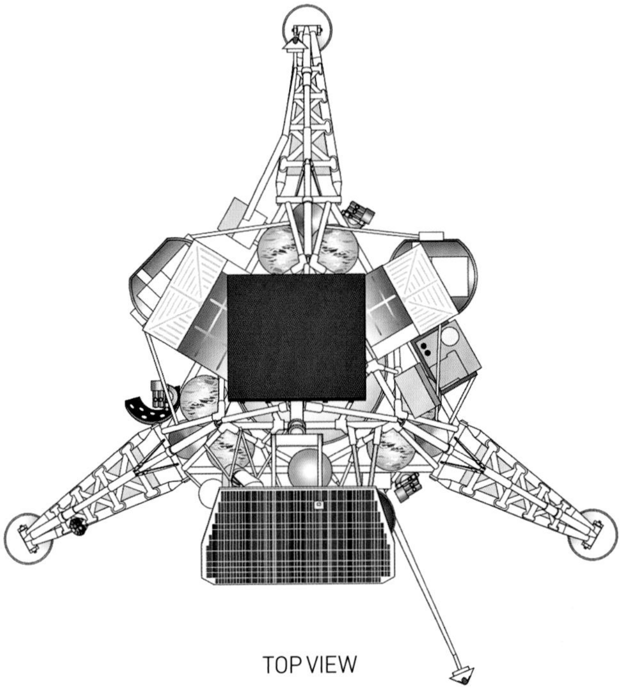

TOP VIEW

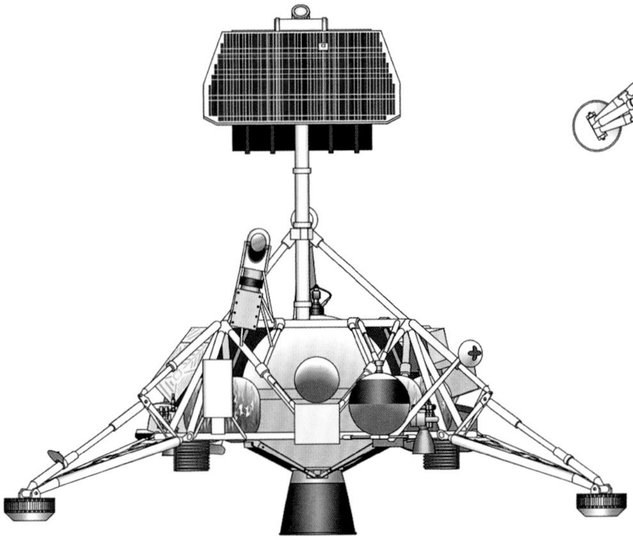

FRONT VIEW

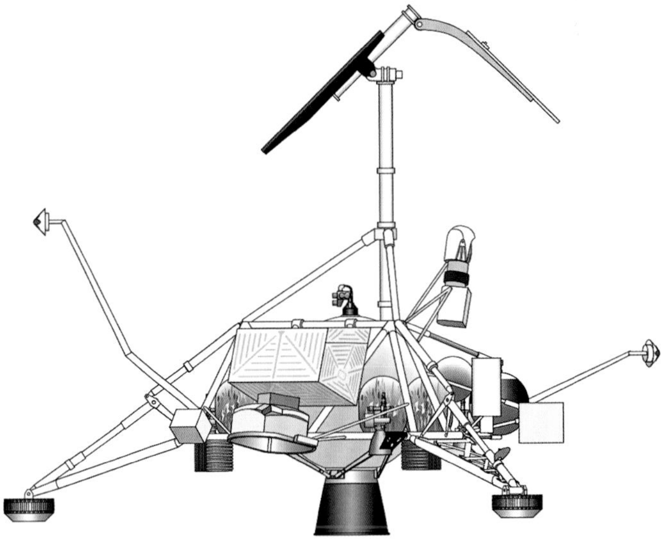

SIDE VIEW

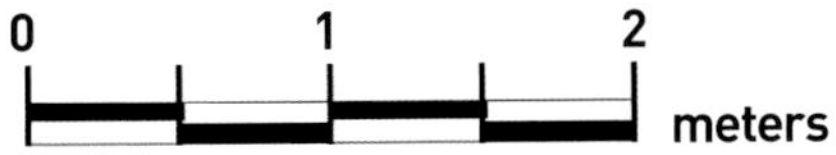

0   1   2

meters

Mariner 9 Probe

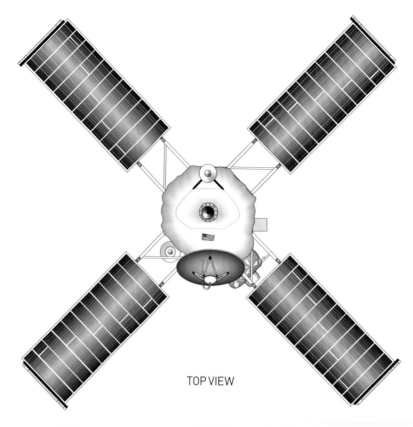

TOP VIEW

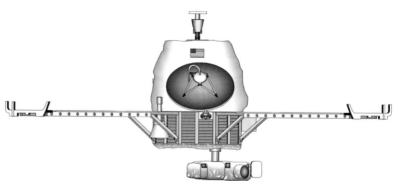

FRONT VIEW

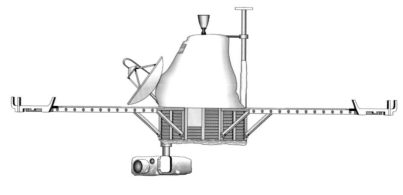

SIDE VIEW

0   1   2

meters

more ambitious journeys, this time to Mars. They abandoned the Ranger-based platform of Mariners 1 and 2 and designed a completely different spacecraft that weighed substantially more (570 pounds, 260 kilograms each). Mariners 3 and 4 not only took pictures like Mariner 2, but could also detect cosmic rays, ionization, magnet fields, radiation, and other features in the Martian environment.

But once again, Mariner experienced problems. After launching in November 1964, Mariner 3 disappeared from ground control when its solar arrays stayed shut. Mariner 4, however, enabled Mars to become a more comprehensible place. As it passed the Red Planet on July 14 and 15, 1965, it took twenty-one photographs of a barren, cratered landscape, and recorded a surface temperature of -148 Fahrenheit (-100 Celsius) with a thin atmosphere. Scientists reluctantly concluded that life of any kind, at any time, seemed highly unlikely.

The next batch of Mariners gave JPL more reason for hope. Mariner 5 flew by Venus in October 1967 and detected an atmosphere ninety times denser than that of Earth. Mariners 6 and 7 reached Mars almost concurrently, in July and August 1969, on a close flyby mission to penetrate its atmosphere and assess its surface. This encounter left researchers slightly more optimistic that life may have once existed there; the paired Mariners found large concentrations of carbon dioxide and water in the environment. But Mariner 8 returned the program to its funk. Launched in May 1971 to orbit Mars, it fell to Earth when the Centaur stage of the Atlas rocket failed just after separation.

Out of the ashes of Mariner 8, two final Mariners proved to be not just fruitful in themselves, but inaugurated a golden age of planetary exploration that came into its own as the Apollo program came to an end.

Mariner 9 shared almost nothing with its predecessors except for its exploration of Mars. Four times the weight (at nearly 2,200 pounds/998 kilograms) of Mariners 3 and 4, it represented a quantum leap in spacecraft complexity. It contained an imaging system, ultraviolet and infrared spectrometers, and an infrared radiometer. It reached Mars on November 14, 1971, and became the first spacecraft to orbit another planet. Its mission planners hoped that it would circle Mars long enough to photograph 70 percent of its surface. Once some heavy Martian winds and dust subsided, planetary scientists marveled at the incoming data. As Mariner 9 passed over and over the Martian terrain, what had been obscured now became clear—that, indeed, life may have once existed on the planet. Researchers saw clearly that water once flowed, and that the poles contained water, making past life a possibility. When NASA ended Mariner 9's career in

October 1972, it had mapped 85 percent of Mars and captured 7,329 pictures at a resolution of half a mile to a mile, leaving no doubt about what had been seen.

At first, Mariner 10 seemed almost like a lost cause. When JPL's leaders appealed to NASA headquarters in 1969 to support a mission to Mercury, the planet closest to the sun, their bosses balked, as did some members of Congress. Why send a probe to a "dead" body just as the excitement of the Apollo program reached its climax? The Mariner team had an answer. The year 1973 offered an alignment of planets that enabled a probe to travel around Venus and then, by gravity assistance, to be catapulted to Mercury.

Despite the concerns about its value, JPL got the go-ahead and the Mariner 10 mission proved to be historic. Mariner 10 looked strange, like a creature from Jurassic Park: a dinosaur-era bird of prey with a broad wingspan, a small round body, two closely set eyes, and two long limbs. It consisted of an octagonal magnesium structure that housed its instruments, solar arrays measuring more than 26 feet (8 meters) across, a dual lens television camera, and two booms: one for a low-gain antenna, the other for twin magnetometers. True to its avian features, at 1,109 pounds (503 kilograms), it weighed less than half of Mariner 9. Despite its lightness, Mariner 10 carried a full suite of instruments: an infrared radiometer, ultraviolet spectrometers, plasma detectors, charged particle telescopes, and magnetometers.

Mariner 10 became the first spacecraft to visit two planets, the first to attempt planetary gravity assistance, and the first to come back to a planet after an initial encounter. The calculations that returned Mariner 10 to Mercury grew out of the fertile mind of Italian mathematician Giuseppe Colombo, who predicted that once the spacecraft passed Mercury it would enter a 176-day solar orbit, at the end of which—with only minor course corrections and minimal propulsion—it would appear again at the innermost planet. After its launch on November 3, 1973, Mariner 10 flew by Venus in February 1974, took 4,100 pictures, and sped off to Mercury, where on its closest encounter in March 1974 it flew as close as 437 miles (703 kilometers) from the surface and made intensive scientific measurements of its barren, moonlike topography. It then flew two loops around the sun, each time returning to observe Mercury, until NASA decommissioned it in March 1975.

Pioneer 10

Among all American spacecraft, the Pioneer series represented the first to travel beyond the inner planets, to pass the gaseous giants at the far reaches of the solar system, and to enter interstellar space. As such, they embody a quantum advance in robotic capability.

They began in 1958 with Pioneer 3. A tiny vehicle that weighed just 13 pounds, it failed in its attempt at a lunar flyby. The following year Pioneer 4 became the first US probe to escape Earth's gravity, but it overshot the moon and returned no pictures. After these mishaps, the Pioneers turned to the sun. In 1960, Pioneer 5—with a mass of only 95 pounds—set off on a voyage around our familiar star, followed five years later by Pioneer 6, an on-orbit weather station that predicted solar storms (enabling government and business customers to protect their electronics systems from disruption). Then, from 1965 to 1968, NASA launched Pioneers 7, 8, and 9, a constellation designed to measure the influence of solar flares on the Earth.

What came next began a radically new chapter in space exploration. As early as 1967, the planetary fraternity at NASA considered a mission to Jupiter, the titan of the solar system with a mass more than twice that of all the other planets combined. The space agency decided to stick with success, giving primary control of the project to the same team that produced Pioneers 6 to 9: NASA Ames Research Center in Sunnyvale, California, and its industry partner, the Thompson Ramo Wooldridge Company (TRW). Problematic as the attempt at long distance travel from the Earth may have been, perhaps the greater difficulty involved the distance from the sun: a gulf of 484 million miles (779 kilometers)—too far away for solar panels to be of any use as a source of power. To compensate, the US Atomic Energy Commission supplied NASA with the nuclear materials and equipment necessary to generate heat, which the TRW engineers transformed into Pioneer's source of electricity.

The space agency decided to build two final, twin Pioneers and send them in different directions. These spacecraft resembled a three-legged sea creature, with a concave dish simulating the body and three protruding antennae in place of legs. Between its extremities (from the medium-gain to the omni-directional antenna), the spacecraft measured 9.5 feet (2.9 meters) long; its widest cross-wise dimension (on the high gain dish antenna) measured 9 feet (2.7 meters). Mounted below and parallel to the dish antenna, a platform held Pioneer's many science instruments. It weighed 571 pounds (259 kilograms).

Launched on March 2, 1972, Pioneer 10—boosted by the Atlas-Centaur rocket combination—flew at 32,188 miles/51,800 kilometers per hour, a record speed for any spacecraft to that date. As the first probe to fly beyond Mars, Pioneer 10 also became the first to pass through the asteroid belt between the Red Planet and Jupiter, and its project managers cheered when it cleared this danger zone with only minor damage. It arrived at

Jupiter in November 1973, sent back three hundred pictures, came closest to the mammoth planet the next month, then flew off toward Saturn. It crossed the orbit of Neptune in June 1983, after which it became the first object fashioned by humans to leave the planetary solar system. NASA ended practical contact with Pioneer 10 in March 1997 and received a last, faint signal in January 2003. In all likelihood, it still flies on in the twenty-first century in interstellar space.

Then came the last of the Pioneers. Pioneer 11 left the Earth on April 15, 1973, flying much faster (106,000 miles/171,000 kilometers per hour) even than its brother spacecraft. At its closest encounter in December 1974, it got three times closer to Jupiter (26,600 miles/42,809 kilometers) than Pioneer 10, enabling accurate measurements of the severe Jovian radiation. It crossed Saturn's ring plane in September 1979, captured images of its flat terrain from as low as 13,000 miles (20,900 kilometers), and recorded the planet's bitter temperatures (averaging -290 degrees F/-180 C). It finally approached Neptune in February 1990, whose orbit it crossed in the opposite direction to Pioneer 10. Pioneer 11 then followed a path beyond the solar system, on which it probably continues until today. NASA retired Pioneer 11 in September 1995 after its power source dropped below levels required for contact.

Assuming the viability of both, roughly forty-five years into their journeys Pioneers 10 and 11 have traveled, respectively, about 10 and 8.5 billion miles from Earth.

If sentient beings encounter either of them, the last Pioneers both carry simple pictographic representations of humanity etched on 9-inch (229-millimeter) by 6-inch (152-millimeter) aluminum plaques. They show stark, primitive images: the male and female anatomy, the planets of the solar system and Earth's position in it, and the date that the spacecraft left the Earth.

Viking 1 and 2

As soon as they saw the pictures sent home by Mariner 4 during the mid-1960s, scientists at JPL realized how little they really knew about the Red Planet. Although Mariners 6, 7, and even 9 revealed more complexity, there remained a gnawing sense of incompleteness. The urge to learn more, coupled with the knowledge that the USSR had been trying to enter the Mars sweepstakes (so far unsuccessfully), motivated NASA to consider a landing on Mars. As funding for the Apollo program began to diminish even before the first moonwalk, the JPL researchers and others at NASA dreamed of shifting the nation's attention to a big, bold, Mars project, which they called Voyager Mars. Voyager Mars grew in proportion to the vehicle required to launch it; with no Saturn 1s in production, mission

Pioneer 10 Probe

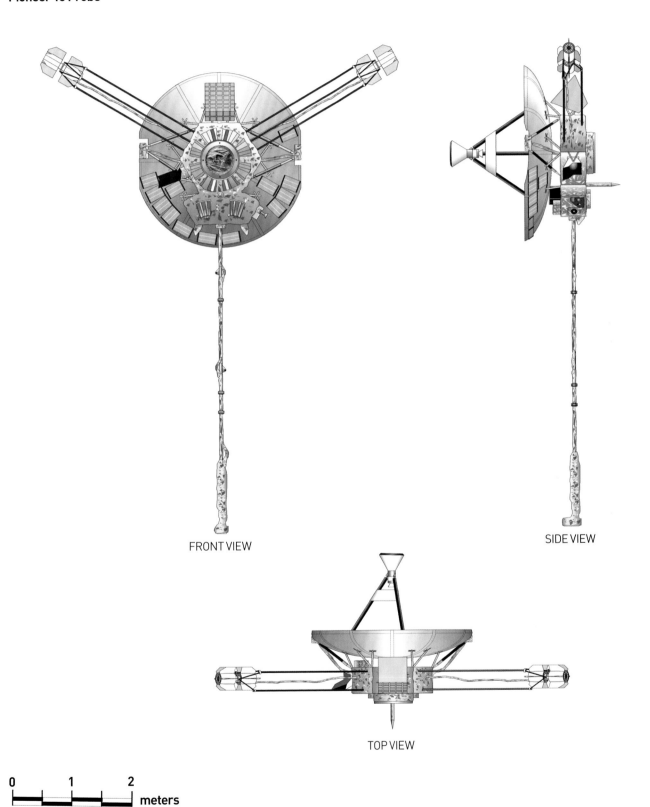

FRONT VIEW

SIDE VIEW

TOP VIEW

0   1   2   meters

planners asked for a Saturn V—and for a $5 billion budget. But the US Congress thought differently, concluding that the money saved from Apollo's decline might be used for purposes other than a massive space robotics mission, and rejected Voyager Mars.

In the wake of this setback, the space agency selected the NASA Langley Research Center to lead a new, scaled-down project named Viking. A group of humbled NASA officials testified before Congress, this time promising two identical Mars spacecraft for $750 million. They won approval, but encountered two years of delay due to Nixon administration budget cuts. During this period, the Viking team scheduled the launch of the first mission for 1975 (a year in which planetary alignments at least partially favored a return to Mars).

In the meantime, Langley chose the Martin Marietta Company to fabricate the lander, JPL to build the orbiter (applying its wealth of Mariner experience), and NASA Lewis Research Center to develop the Titan III-Centaur launch vehicle. Program officials also persuaded Congress to accept a budget increase to $830 million.

The resulting machine looked impressive but slightly awkward, with a broad clamshell on top (containing the lander) and an inverted cone underneath (with the orbiter).

The JPL designers fashioned the Viking orbiter in the mold of Mariner 9, only bigger, heavier (5,157 pounds/2,339 kilograms), and with advanced imaging and communications systems. The lander (1,270 pounds/576 kilograms) proved far more problematic. NASA insisted that Martin Marietta furnish them not just with full-color photography equipment, but also with twin cameras for stereo depth perception. Additionally, the chromatograph-spectrometer used to detect organic substances needed to be the state of the art. One program official confessed, "It was soon apparent that we had bitten off more than we could chew."

Despite some last-minute corrections to Viking's computer hardware, Viking 1 lifted off on August 20, 1975, and reached Mars on June 19, 1976. Soon, however, clear images from the orbiting spacecraft made it obvious that the terrain chosen for the landing lacked the necessary smoothness, so mission control redirected it to a more favorable area. After about a month of searching, the lander finally separated from the orbiter and, following a complex series of maneuvers, made a safe touchdown.

The data streamed in. The good news: the color photos proved to be astounding, and Viking 1 transmitted periodic weather updates, detecting low temperatures of -123 degrees F (-86 C) at dawn and -27 degrees F (-33 C) in the afternoon. The bad news: even the simplest forms of life eluded Viking's sensors. The lander's robot arm succeeded in collecting a sample for its

biological laboratory, and while some of the data hinted at the possibility of life forms in the distant past, no organic compounds came to light.

A month after Viking 1 left on its mission, Viking 2 set off for the Red Planet on September 9, 1975, and entered orbit around Mars on August 7, 1976. As before, the landing site needed to be modified based on visual evidence of excessive ruggedness, so the lander—which again touched down perfectly—made contact near the edge of Mars's polar ice cap. Despite the continued hope for the evidence of former life, the scoop of materials analyzed on this trip also returned indefinite results. On the positive side, the Viking orbiters collectively photographed 97 percent of the Martian surface, yielding 51,500 images—seven times the number of Mariner 9, and this time in depth and in color. But the question of organic activity on Mars remained unresolved.

Voyager 2

Although President John F. Kennedy initiated the race to the moon, Richard M. Nixon surpassed all other presidents in showing a sustained interest in American space policy. Indeed, Nixon's core objectives—lowering launch costs, developing international cooperation, and pursuing planetary exploration— became fundamental to NASA policy.

Scientists at the Jet Propulsion Laboratory understood that a quirk of nature conformed to one of Nixon's key space preferences. They realized that from 1976 to 1979, the outer planets of the solar system would be in an alignment seen once every 175 years, enabling spacecraft to move by gravity-assist (gaining speed as they flew by each planet), thus reducing travel time and eliminating the need for heavy onboard propulsion. The JPL team won over NASA Administrator Thomas Paine to the project, and he in turn persuaded congressional and White House officials to support what became known as the Grand Tour—that is, to fly past all the gas giants of the solar system: Jupiter, Saturn, Uranus, and Neptune, in addition to many of their moons.

In contrast to the final two Pioneers, where NASA's Ames Research Center contracted with private industry (TRW) to build the spacecraft, the Grand Tour vehicles proposed by JPL would not only be designed, but fabricated on the JPL campus. Also unlike the Pioneers, these would not be a series of spacecraft deployed over many years, but just one pair of identical twins, launched in quick succession. Called Voyager 1 and 2, JPL patterned them after the lab's successful planetary probes from the early 1970s, Mariners 9 and 10.

These powerful but relatively small vehicles looked buglike, with bodies in the form of a high-gain dish antenna, and five booms and antennas projecting outward, like ungainly legs.

**Viking 2 Probe
Prelanding Configuration**

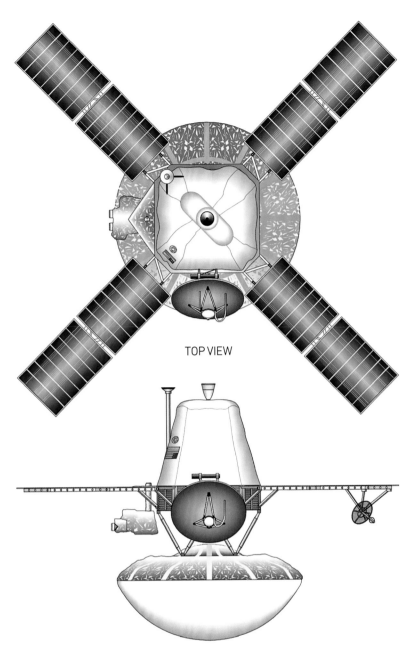

TOP VIEW

FRONT VIEW

LANDER VIEW

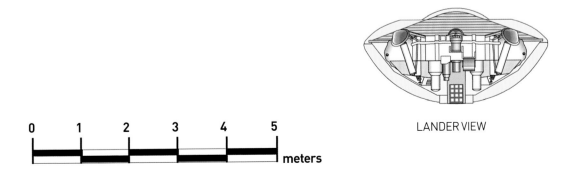

0　1　2　3　4　5

meters

Voyager 2

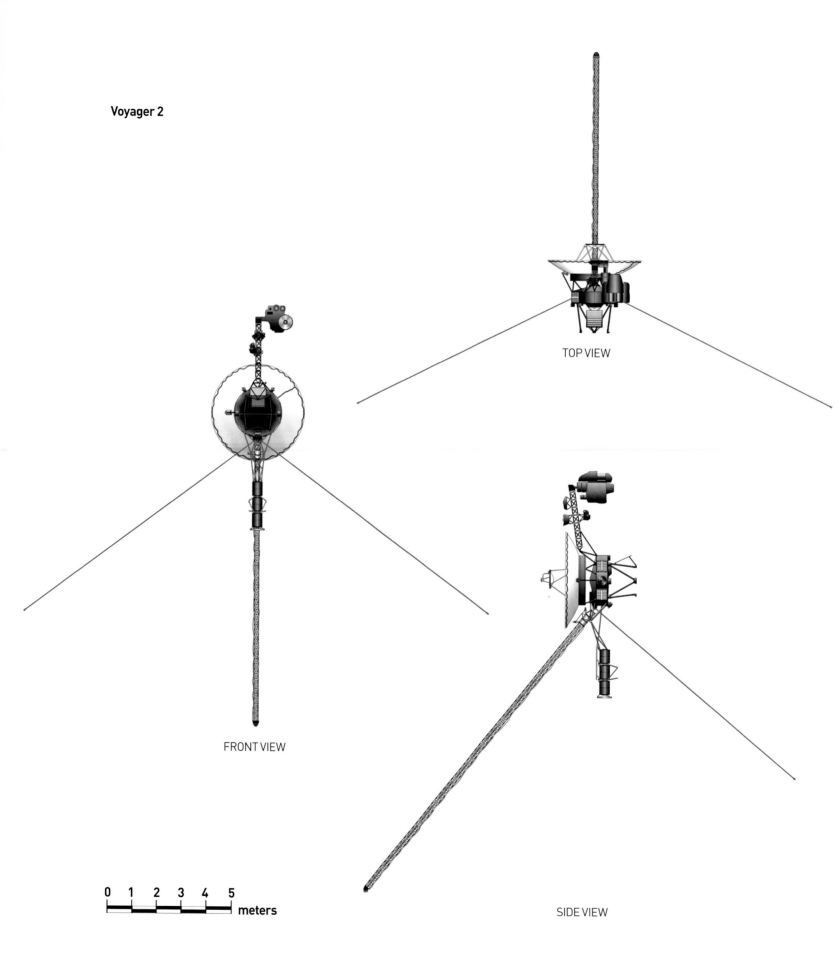

TOP VIEW

FRONT VIEW

SIDE VIEW

0 1 2 3 4 5 meters

The electronic bus, located in a ten-sided container, measured just 1.5 feet (.47 centimeters) tall; a gap of only 6 feet (1.8 meters) separated the spacecraft's most prominent flat surfaces. The longest booms included the 8.2-foot (2.5-meter) one for science instruments, and the 42.6-foot (13-meter) one for magnetometers. The Voyagers also carried two whip antennas, each 33 feet (10 meters) long. The spacecraft weighed 1,820 pounds (823 kilograms).

Voyager 2 flew first and found new worlds on its long-distance trek. Launched on August 20, 1977, aboard a Titan III-Centaur rocket combination, by July 1979 it made its closest approach to Jupiter, meanwhile observing its moons Amalthea, Io, Callisto, Europa, and Ganymede. It sent home superb photographs and motion pictures of the swirling Jovian atmosphere. Then it encountered Saturn in August 1981, photographing its rings, as well as its moons Hyperion, Enceladus, Tethys, and Phoebe. In January 1986, Voyager 2 passed Uranus, where, astonishingly, it discovered ten new moons. Finally, the first ever flyby of Neptune occurred in August 1989. The spacecraft flew to within 2,800 miles (4,506 kilometers) of its surface, finding five new moons, a hydrogen/methane atmosphere, and winds up to 680 miles per hour (1094 kilometers per hour).

Voyager 1 lifted off after its sister ship on September 5, 1977, but reached Jupiter (March 1979) and Saturn (November 1980) sooner because it took a more direct pathway. It alerted astronomers to a Jovian ring in 1979 and flew close by another of Saturn's moons, Titan.

Both spacecraft began to escape the solar system in 1989, at which time NASA called their journeys the Voyager Interstellar Mission (VIM). During this phase, the JPL team hoped to learn about space at the limits of the sun's influence. The scientists wanted to define the boundaries of the heliopause—the outer edge of the sun's magnetic field—where interstellar space begins, characterized partly by solar winds that decline from supersonic to subsonic speeds.

Barring the unforeseen, the Voyagers' science instruments will function fully until about 2020, after which time project engineers will gradually shut them down to conserve electrical power until, by 2025, all experimental work will end. Beyond that, an engineering-only mission phase may go on until 2036.

Roughly forty years after they lifted off, Voyager 1, the most distant object fashioned by human beings, has traveled roughly 12.7 billion miles, and Voyager 2, 10.4 billion miles (20.4 and 16.7 billion kilometers, respectively). At these positions, Voyager 1 now flies in interstellar space and Voyager 2 progresses through the heliosheath (the outer layer of the heliosphere, where the slowing of solar wind occurs).

Like Pioneers 10 and 11, they both carry greetings for any reasoning beings that may encounter them as they plunge deeper into the Milky Way. Each holds a gold-plated, 12-inch phonograph record whose contents—guided by the famed twentieth-century astronomer and science popularizer Carl Sagan of Cornell University—differed greatly from the simple etched symbols of the Pioneers. These newer time capsules reflect a far broader appreciation of the human experience on Earth, containing samples not only of images, music, and spoken languages, but of elemental sounds: a horse and cart, a chimpanzee, Morse Code, human laughter, and a heartbeat.

Sputnik 1

Despite the space race that eventually enveloped the United States and the USSR, Sputnik 1—the first man-made object to orbit the Earth—began in an atmosphere of scientific inquiry, untainted by politics. Events that led to the Soviet satellite went into motion at the start of the Eisenhower administration, when the International Council of Scientific Unions declared 1957–1958 to be the International Geophysical Year (IGY). IGY organizers urged interested governments to pursue their research not just with the available sounding rocket payloads that entered space briefly, but with orbiting spacecraft. Of course, satellites required powerful launch vehicles, and since the United States and USSR both pursued advanced ballistic missile programs, they saw the prospect of success and agreed to participate.

In a technical sense, Sputnik 1 began well before the call of IGY, with the research pursued by Mikhail Tikhonravov, a founding figure in the Soviet space program, and his associates at the NII-4 military institute. Tikhonravov's group pioneered work in artificial satellites, from which Sputnik 1 evolved.

The Soviet government gave its formal approval to the IGY challenge in January 1956 and targeted 1957 as the year of launch. As envisioned originally, Sputnik 1 (known then as Object D) weighed between 2,205 to 3,086 pounds (1,000 to 1,400 kilograms), of which 441 to 661 pounds (200–300 kilograms) consisted of scientific cargo. But rethinking in 1956 persuaded Sergei P. Korolev—the preeminent figure of the early Soviet space program and a friend of Tikhonravov—that the complex instruments designated for Object D could not be prepared on the short schedule of IGY. So the mission planners got orders to fall back to a different model: *prosteishy sputnik*—simplest satellite. Its mass plummeted to between 176 to 220 pounds (80 to 100 kilograms), suggesting a practical choice made by Moscow: to favor timeliness over science, with the IGY goals to be fulfilled later.

The Soviet Union pursued this objective with the single-mindedness possible in a unitary state. In contrast, American

Sputnik (PS-1) Satellite

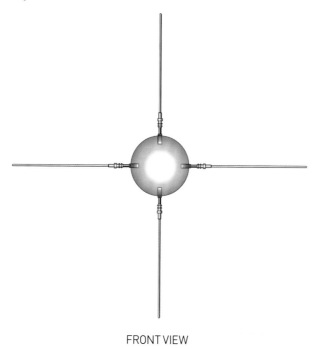

FRONT VIEW

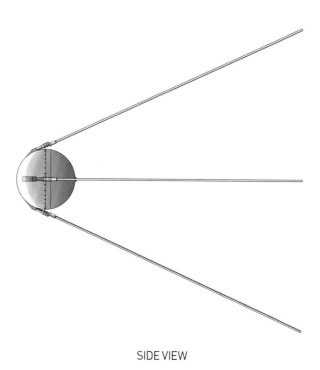

SIDE VIEW

0   1

meters

policy makers, especially President Dwight Eisenhower, preferred not to dilute high-priority military programs such as ballistic missiles with civil programs, and so they divided resources and talent for the IGY mission among competing teams.

Sputnik 1 lifted off from the Baikonur Cosmodrome in Tyuratam, Kazakhstan, on October 4, 1957, aboard an R-7 Semyorka rocket. Announcements in the official Soviet news agencies seemed muted given the magnitude of the achievement, simply reciting the details of the flight. In sharp contrast, the Western media cast the event in capital letters. *The New York Times* ran a story from Moscow dated October 5 under a front page, eight-column, banner headline: "SOVIET FIRES EARTH SATELLITE INTO SPACE; IT IS CIRCLING THE GLOBE AT 18,000 M.P.H.; SPHERE TRACKED IN 4 CROSSINGS OVER U.S." The traditionally understated British Broadcasting Corporation (BBC) suggested in its earliest reporting that the launch heralded a new age, fraught with anxiety: "The satellite's weight has led some American experts to speculate that the rocket which launched it might also be capable of carrying a nuclear weapon thousands of miles. The fact that Sputnik is expected to fly over the US seven times a day has caused unease." And many watched it: visual observers at 150 stations across the world saw it streak across the sky, night after night. Additionally, ham radio operators everywhere listened to its signals at 20.005 and 40.002 megahertz, which beamed for three weeks.

Of course, the BBC's supposition about the R-7 rocket proved to be right, but Sputnik 1 hardly represented a menace. A polished aluminum sphere almost 23 inches (58 centimeters) in diameter, it featured four protruding antennas measuring from 8 to just under 10 feet (2.4 to 2.9 meters) long. The world's first satellite weighed just over 184 pounds (83.6 kilograms), and although light, it contained a few scientific instruments (held in a sealed capsule filled with nitrogen under pressure). Sputnik 1 collected data regarding the density of the upper atmosphere and its transmitters broadcasted radio signals into the ionosphere. It also had the capacity (through temperature fluctuations) to detect meteoroid strikes if any penetrated the spacecraft's skin; none did. Downlink telemetry amassed information about temperatures inside and on the exterior of the satellite.

Sputnik disintegrated upon reentering Earth's atmosphere on January 4, 1958, after 1,440 orbits. Of course, because of Korolev's decision to put the IGY deadline above all other considerations, its main value did not lie in its science, but in its symbolic political importance. It aroused doubts in the minds of some about America's supremacy in science and technology. But even more, it raised questions in those post-colonial countries in the process of choosing their political structures whether the United States or the Soviet Union offered the more effective model of ideology and government. And it caused a partisan firestorm in the United States, resulting on October 1, 1958, in the establishment of the National Aeronautics and Space Administration—three days before the first anniversary of Sputnik 1's launch.

Sputnik 2

Shocking as the launch of Sputnik 1 may have been to Western observers, the liftoff of Sputnik 2 just a month later may have unleashed an even a bigger jolt. Sputnik 1 offered the element of surprise, but the satellite itself had limited scientific value (a fact admitted by the Soviets in their eagerness to strike first in the IGY sweepstakes). On the other hand, Sputnik 2 qualified as a full-scale success by any standard, including those of the IGY. Moreover, it followed Sputnik 1 so closely that no one could call the Soviet successes accidental.

In the runup to Sputnik 1, it dawned on many in the Soviet space hierarchy that the deadline of 1957 to launch an IGY satellite left no margin for a fully realized science spacecraft, then on the drawing board as Object D. So Sputnik 1—the "simple satellite," as the Soviet designers called it—went into orbit instead. That raised the question of the role of Sputnik 2. Object D still could not be prepared in time, and sending up another Sputnik 1 lacked headline appeal. Into the breach stepped Sergei Korolev. He realized that extra weight rested on the Sputnik 2 decision because Soviet premier Nikita Khrushchev wanted the second Soviet satellite to celebrate the fortieth anniversary of the Bolshevik Revolution.

Korolev found the key to the dilemma in the past. During the early 1950s, he and his colleagues had flown dogs on high-altitude missions on board ballistic missiles. In 1955, Korolev extrapolated these experiences and began preparations to launch a canine passenger on an around-the-world flight. The concept received approval in 1956 with the argument that sending a living creature into orbit represented a necessary step to proving the safety of spaceflight for human beings. But the actual go-ahead did not occur until October 10 or 12, 1957, leaving less than a month for final plans and to satisfy Khrushchev's wish.

Amid this rush of activity, at least Korolev's team did not need to concern themselves with rocketry: Sputnik 2 flew into space on a modified version of the R-7 used in Sputnik 1. But everything else changed radically. In contrast to Sputnik 1's 23-inch (58-centimeter) aluminum ball that weighed only 184 pounds (83.6 kilograms), Sputnik 2 looked (and functioned) like an object

Sputnik 2 (PS-2) Satellite

FRONT VIEW

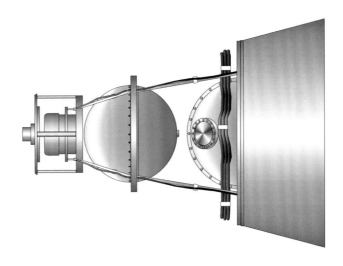

SIDE VIEW

0   1
meters

from a future age. Its tall conical design measured 13 feet (4 meters) high, 6.6 feet (2 meters) at the base, and it weighed 1,121 pounds (508.3 kilograms). Sputnik 2 held several compartments, containing scientific instruments, radio transmitters, a programming unit, a telemetry system, and, most importantly, a sealed cabin with air regeneration and temperature control for its four-legged traveler. In addition, two spectrophotometers measured ultraviolet, x-ray, and cosmic ray emissions.

Thirty-two days after Sputnik 1's launch, Sputnik 2 lifted off from Baikonur Cosmodrome, Tyuratam, Kazakhstan, on November 3, 1957. Once in orbit, Laika ("Barker")—part husky, part terrier, and selected for her good temperament from among ten candidates—could sit or stand, had access to gelatinized food and water, but had to be chained and fitted with a harness. Electrodes monitored her heart rate and other bodily functions. According to the biometric telemetry, she survived the launch well. The Soviet press reported at the time that Laika lived for seven days, and although her vital signs appeared normal for three orbits, high temperatures inside the cabin (due to loss of thermal insulation) probably ended her life in one or two days. Regardless of the actual events, mission planners knew that she would not survive reentry.

Sputnik 2 continued to orbit the Earth until April 14, 1958, when it burned up in the atmosphere, by which time the United States had answered the Sputnik successes with Explorer 1 and Vanguard 1 (launched, respectively, on January 31 and March 17, 1958).

Sputnik 3

With the launch of Sputnik 3 in spring 1958, its prime mover— Sergei P. Korolev—finally achieved the full scientific satellite that Russia intended in response to the International Geophysical Year call for orbital flight. Korolev had no choice but to postpone

Sputnik 3 because of its advanced science payload and telemetry. Instead, the Soviets relied on Sputniks 1 and 2 to represent them in the earliest days of the space race.

Korolev paired Sputnik 3 with the R-7 rocket, modified from the version that lifted Sputniks 1 and 2 into orbit. Compared to the preliminary Object D of 1956, this one proved to be even more ambitious: an automated scientific laboratory, it included two cosmic ray detectors, in addition to twelve scientific instruments that measured such phenomena as meteorite activity, solar radiation, electric and magnetic fields, electrical charges, positive ion concentration, and the pressure and makeup of the ionosphere.

Unfortunately, the creators of Object D had to wait a little while longer before realizing their objective. On April 27, 1958, Sputnik 3, atop the R-7 Semyorka, rose from the launch pad at Baikonur in Tyuratam, Kazakhstan, and for the first minute and a half, everything seemed normal. But in that moment, the structure disintegrated and fell to the ground in flaming pieces. Korolev informed Moscow that at ninety seconds, vibrations onboard the rocket brought down the mission. He gave orders that next time, the R-7's engines needed to be throttled back starting at the eighty-fifth second. Meanwhile, security officers confiscated all recordings of the launch, and the accident remained unreported for decades.

Better news occurred during the second attempt on May 15, 1958, when Sputnik 3 went into orbit. Most of its instruments functioned for more than two weeks (although consequentially, the mission's tape recorder failed). Sputnik 3 stayed aloft for 692 days—twice the projected length of time—when it reentered the atmosphere on April 6, 1960.

Korolev debriefed Object D personnel about the results in August 1958. He told them that success in Sputnik 3 presaged bold ventures to come, involving lunar probes and manned circumnavigations of the moon. Sputnik 3, he said, gained plenty of scientific information—with limitations. It detected the outer radiation belts of the Earth and sent home details about the composition of the upper atmosphere, meteoric particles, magnetic and electrostatic fields, heavy nuclei in cosmic rays, photons in cosmic rays, and the composition of charged particles. Yet, because of the mishap with the tape recorder, Sputnik 3 could only return data in real time, therefore confining its collection solely to Soviet tracking stations as the spacecraft flew over the USSR.

Just as Sputnik 2 looked different from Sputnik 1, Sputnik 3 bore almost no resemblance to its predecessors. Sixteen times heavier than Sputnik 1 and two and a half times the weight of Sputnik 2, Sputnik 3 carried a payload with a mass more than twice that of the five previous Soviet and American orbital

Sputnik 3 (Object D-2) Satellite

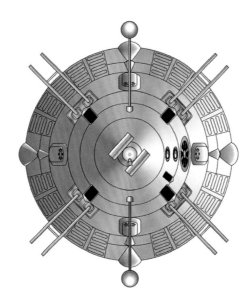

FRONT VIEW

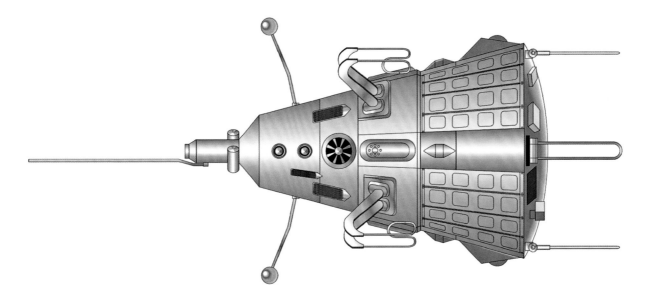

SIDE VIEW

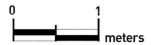

0 1 meters

Lune 3 (Ye-2A) Probe

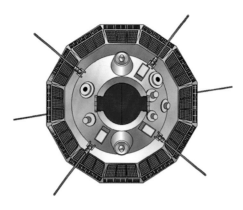

FRONT VIEW

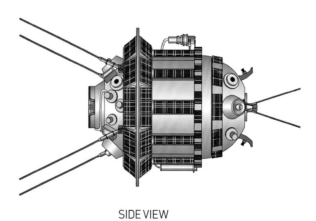

SIDE VIEW

0   1

meters

spacecraft combined. Sputnik 3 also distinguished itself as the biggest satellite to date. And it looked futuristic—not accidentally, more like a like a capsule than a probe. It measured almost 12 feet (3.6 meters) long, nearly 5.6 feet (1.7 meters) in diameter, and weighed about 2,926 pounds (1,327 kilograms).

Sputnik 3 catapulted the USSR well ahead of the United States in space. Even though three successful American launches followed Sputnik 2 (Explorer 1, Vanguard 1, and Explorer 3), the United States also experienced four failures (three Vanguards and Explorer 2) during the same period (December 6, 1957, to April 29, 1958). And although Explorer 3 proved the existence of the Van Allen radiation belts, at just 31 pounds (14.1 kilograms), it posed no match for the giant Sputnik 3's elaborate science payload.

Luna 3

At the same time the top Soviet space authorities conceived of Sputnik, they also planned for the exploration of the moon. The Soviets tried to reach the moon as early as September 1958, then in October, and again in December—all without success. Finally, on January 2, 1959, they launched Luna 1, the first spacecraft ever to fly beyond the Earth's orbit. Only Sputniks 1, 2, and 3 preceded Luna 1 in space. And nearly three years passed between the first Soviet moon shot and that of Ranger 1, America's initial lunar explorer.

But Luna 1 did not behave as planned. Designed to strike the moon, it instead passed to within 3.7 miles (5.95 kilometers) before veering off into an orbit around the sun. Luna 2, the second lunar impactor, fared much better. Launched on September 12, 1959, it crashed into the Sea of Serenity, making it the first spacecraft to land on another planetary body.

Luna 3, which lifted off from the Baikonur Cosmodrome on October 4, 1959, proved to be the champion that Moscow wanted. Small and simple at just 4.26 feet (1.3 meters) long and 3.9 feet (1.2 meters) at maximum diameter, it weighed 614 pounds (278.5 kilograms). With a broad canister at its midsection, Luna 3 had two shallow hemispheres at either end. The probe carried micrometeoroid and cosmic ray detectors on its exterior, and Yenisey-2 cameras, film processing equipment, a radio, batteries, gyroscopes, propulsion systems, and other gear inside.

Luna 3 followed a roundabout route to its target. Its ground controllers first placed it in an elliptical Earth orbit. As it passed behind the moon and circled back toward home, lunar gravity changed its trajectory so that, without the need for midcourse corrections, Luna 3 fell into lunar orbit. This maneuver represented the first use of gravity-assist by any spacecraft. While revolving around the moon, Luna 3 came as close as 3,853 miles (6,200 kilometers) as it approached the lunar

south pole. Then, flying back around to the far side, the sunlight triggered its cameras, which operated for forty minutes, taking twenty-nine images that covered 70 percent of the surface. On its homeward trip, the spacecraft transmitted seventeen grainy but comprehensible photographs, including a composite view of the far hemisphere, as well as pictures of two large seas, subsequently named the Mare Moscovrae (Sea of Moscow) and Mare Desiderii (Sea of Dreams).

Luna 3 represented the first of many long-lived Soviet spacecraft series: indeed, the Luna family continued from 1959 to 1976. Its missions persisted despite some failures, and the Luna made historic contributions to mankind's understanding of Earth's celestial neighbor.

Luna 4-9

After the success of Luna 3, the Soviet Union entered a new phase of its moon program by redesigning its lunar explorers for soft landings. American intelligence agencies correctly assumed that these missions prefigured future attempts at a manned landing or even a permanent base. To realize these objectives, the famous Soviet aerospace designer Sergei Korolev worked single-mindedly to catch and surpass the Americans in their ambition to send humans to the lunar surface.

But as Korolev's engineers discovered to their frustration, Luna 3 represented the last success for a long time. In fact, in an incredible streak of mischance, not one of thirteen Luna missions from April 1960 to December 1965 succeeded. Five of the mishaps occurred because of the spacecraft: Luna 4, launched in April 1963, flew past the moon when it missed a course correction; Luna 5 failed to decelerate and impacted the moon in May 1965; Luna 6 also experienced course correction problems and bypassed the moon in June 1965; Luna 7's attitude control malfunctioned and it crashed in October 1965; and, in December 1965, Luna 8 suffered the same fate, for the same reason, as Luna 7. Just as disappointing, during the same timeframe, eight launch failures happened when the Luna, Molinya, Molniya-M, and Molniya-L boosters either did not lift their payloads into space or stranded them in low Earth orbit.

In part, these losses can be explained by the complexity of landing spacecraft intact on distant planetary bodies. Indeed, the machines sent on these travels differed drastically from Luna 3 in size, weight, design, and sophistication. Lunas 4 to 9 measured almost 9 feet (2.7 meters) long and weighed 3391 pounds (1,538 kilograms)—more than twice the length and five times heavier than Luna 3. The newer models looked nothing like the Luna 3 cylinder, but instead resembled an octopus without tentacles: a spherical lander on top connected to a somewhat

Luna 9 (Ye-6) Probe

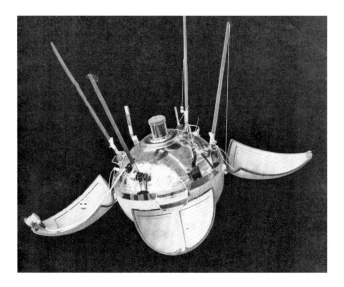

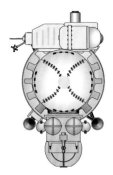

FRONT VIEW

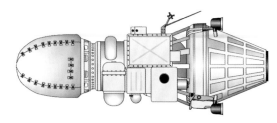

TOP VIEW

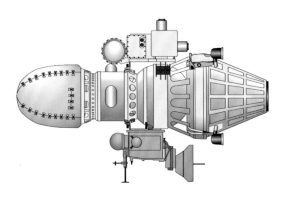

SIDE VIEW

narrow-waisted midsection, attached to an inverted cone at the bottom. The science package inside of the lander consisted of a lightweight panoramic camera (mounted to a mirror that enabled 360-degree photographic coverage) and a radiation detector.

After so many trials and miscues, Luna 9 proved to be the salvation of the Soviet lunar landing program. It left the Baikonur, Kazakhstan, facility on January 31, and arrived at the moon on February 3, 1966. At about 16 feet (5 meters) from the ground, the spacecraft ejected the lander, which bounced several times and came to rest in the Ocean of Storms. About four hours later, the lander's shell, consisting of four leaflike sections, opened outwardly until they touched the surface, providing stability to the instruments within. In all, Luna 9's camera took four panoramic images, showing rocks in the foreground and the horizon at a distance of about 4,600 feet (1.4 kilometers). It also returned radiation data. After three days of operation, the spacecraft's batteries expired and the mission ended.

Despite the many years of misadventure, Luna 9 still defeated its rival, the American Surveyor 1, by about four months to become the first spacecraft to land safely on a celestial body other than Earth. But the US team attempted to salvage at least a partial victory; NASA claimed that its spacecraft made the first *controlled* soft landing and emphasized that it succeeded on the very first Surveyor flight.

Luna 15, 16, 18, 20

Although it became clear in 1969 that the Soviet space agency would not be able to reach the moon before Apollo, the competitive feelings between the two superpowers did not diminish right away. As the Apollo 11 team made plans for the first walk on the moon, Soviet engineers prepared a third

Luna 16 (Ye-8-5) Probe

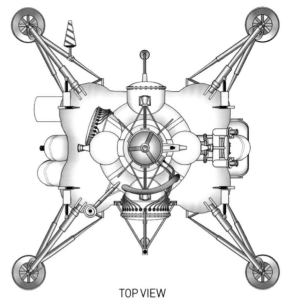

TOP VIEW

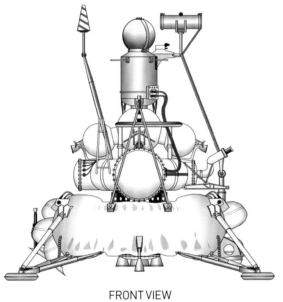

FRONT VIEW

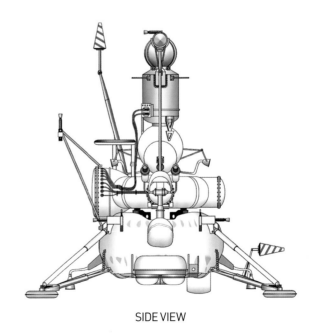

SIDE VIEW

0   1

meters

Luna 17 (Ye-8 N°203) Probe, with Lunokhod 1 Rover
Landing Configuration

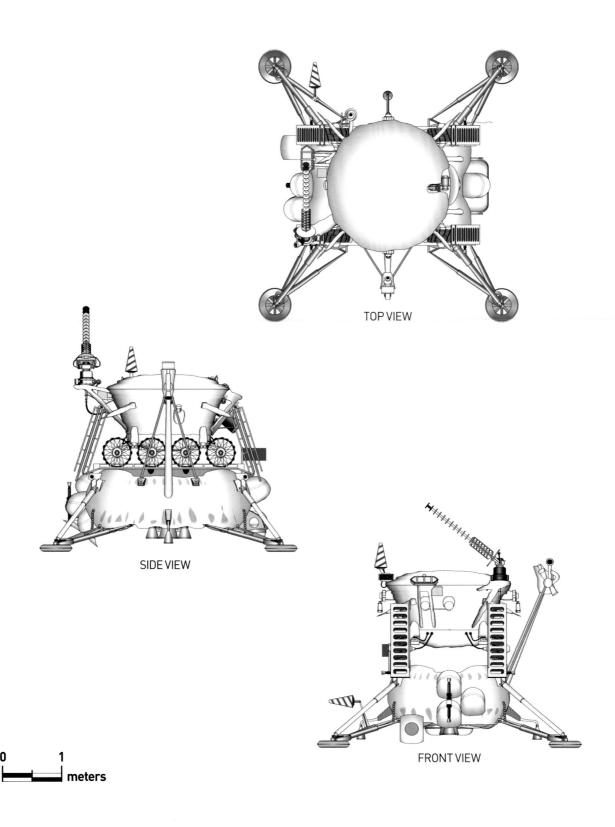

TOP VIEW

SIDE VIEW

FRONT VIEW

0   1
meters

generation of Luna spacecraft for a role that paralleled that of Apollo. Within the Luna program were three different types of spacecraft, each with a specific mission. Luna 3 represented phase one, characterized by orbital observations and intentional crash landings; Luna 9 typified phase two, with soft landings on the surface; and Luna 15 heralded a new age in which Soviet spacecraft succeeded in mining samples of lunar soil and transporting them back to Earth. So, in essence, Luna 15, 16, 18, and 20 attempted to do robotically what Neil Armstrong and Buzz Aldrin accomplished by hand.

Beginning with Luna 15, the physical characteristics of this enduring series of spacecraft again changed radically. This new incarnation looked like a sprawling pyramid of tubes and tanks supported on four splayed legs. It actually consisted of two attached stages, an ascent above and a descent below. The descent stage included a cylindrical body arrayed with fuel tanks, a landing radar, and a main descent engine system (that slowed Luna until it reached a cutoff point, at which time thrust jets handled the final deceleration prior to landing). The descent stage also carried radiation and temperature monitors, telecommunications equipment, a television camera, and, most important for the mission, an extendable arm outfitted with a drill to collect soil samples. The ascent stage, also cylindrical but smaller and topped by a sphere, held a hermetically sealed soil sample container within a reentry capsule, equipped with parachutes for reentry. All these pieces added up to a heavy spacecraft. Luna 15 weighed 12,556 pounds (5,700 kilograms); Luna 16, 12,626 pounds (5727 kilograms); and Luna 18 and 20, 12,346 pounds (5,600 kilograms) each. All traveled into space aboard powerful Proton-K/D rockets.

The Soviets hoped to land and return Luna 15 before Apollo 11 got the chance to carry out one of its main missions—to collect and return lunar samples. It looked like they might succeed. Luna 15 launched on July 13, more than three days before Apollo 11, and went into lunar orbit on July 17. It continued to circumnavigate the moon (in accordance with an agreement with NASA to avoid interference with each other's radio frequencies), during which time Apollo 11 landed on the 20th. Two hours before Neil Armstrong and Buzz Aldrin planned to leave the moon's surface and begin the voyage home, Luna 15 descended toward the Sea of Crises; but it crashed, ending the mission.

Undeterred, the Soviet space agency sent Luna 16 to the moon on September 12, 1970, and on the 20th fired its descent engine. It landed safely, after which ground controllers activated the automated drilling rig, which could penetrate roughly 12 inches (about 30 centimeters) of rock. The drill spilled about 3.5

ounces (100 grams) of lunar soil into a loading hatch at the top of the ascent stage, which ground control sealed. After twenty-six hours on the surface, the ascent stage took off, reentered Earth's atmosphere, and on September 24 a recovery group in Kazakhstan found the soil container. The Soviets made another try, launching Luna 18 on September 2, 1971, but communications ended not long after its descent began, probably due to a crash. Then came another success; Luna 20 left for the moon on February 14, 1972, landed safely in the Sea of Fertility, and used an improved drill to extract 1 ounce (30 grams) of soil from hard rock at a depth of 4 to 6 inches (10 to 15 centimeters). Upon its return, a team discovered the capsule on an island, in a blizzard, in Kazakhstan. Experts dated the sample from Luna 16 at between 3 to 5 billion years old and the one from Luna 20 at roughly 1 billion years older.

The program finally came to an end when Luna 24 went into space on August 9, 1976, eighteen years after the first (unsuccessful) Soviet attempt on the moon. Luna 24 ended positively, making the third successful delivery of lunar soil back to Earth. So, although the Americans enjoyed the crowning successes of Apollo 11, 12, 14, 15, 16, and 17, the Soviets accomplished long-term, systematic robotic explorations of the moon during the 1950s, 1960s, and 1970s, not only with the Luna series, but in November 1970 and January 1973, with the two Lunokhod rovers.

Venera 9–12

In the conflict for supremacy between the United States and USSR during the dawn of the Space Age, what appeared to be one large contest segmented itself into a series of narrow but intense competitions. Who would launch the first satellite? Who would orbit the first man? Who would be first on the moon? Who would launch the first space station?

Among these prominent rivalries, a less-appreciated one took shape early and continued for many years. It involved voyages to the solar system. NASA began its planetary research with Project Mariner (1961 to 1975), which focused primarily, but not entirely, on Mars. But as the United States looked mainly to the Red Planet, the Soviets focused much of their attention on Venus and on developing a spacecraft called Venera (Venus, in Russian). Despite the severe Venutian climate—an average temperature of 864 degrees Fahrenheit, sulfur dioxide clouds rising up to 37 miles (60 kilometers) from the surface, and a thick atmosphere like a heavy smog—the Soviet space agency actually accomplished more on Venus than it did on Mars. From the Soviet viewpoint, even though Mars offered a cooler and less dense environment, Venus held out two key advantages: it avoided a costly duel with NASA and

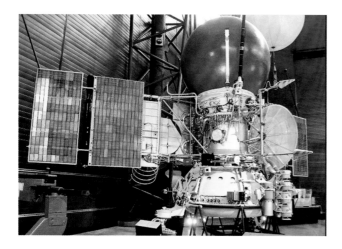

it made for less time-consuming voyages (24 million miles to Venus at its closest point to Earth versus 34 million miles at Mars' nearest approach).

So, rather than compete directly with NASA's missions to Mars, the Soviets concentrated on Venus. They organized their Venera probes much as they did their other big space projects: as long-term investments in money and time, subject to unavoidable mishaps as they progressed. Indeed, eleven successive attempts on Venus (from February 1961 to November 1965) failed to accomplish the objectives of the Soviet space agency. Only with Venera 3 did the Soviets succeed, when the spacecraft crashed (intentionally) on March 1, 1966. Then, favorable events happened in quicker succession: Venera 4 in 1967, and 5 and 6 in 1969. They all sent back data before impacting the planet.

Finally, the controlled (but hard) touchdown of Venera 7 on December 15, 1970—the first planetary landing ever achieved—resulted in a faint transmission of data for twenty-three minutes. Venera 8 followed on July 22, 1972, and sent signals from the surface for fifty minutes.

Venera 9 heralded a new era of Soviet planetary exploration. Through Venera 8, the Soviets relied on a relatively simple spacecraft: a cylindrical main section that housed power systems, guidance, telemetry, the propulsion engine, topped by a globe-shaped descent module. With fuel, Venera 8 weighed 2,600 pounds (1,180 kilograms) and measured just over 6.5 feet (2 meters) tall and 3.3 feet (1 meter) in diameter.

In contrast to the earlier Veneras, the NPO Lavochkin design bureau broke new ground when it developed Venera 9. Oddly, it resembled an eighteenth-century balloon and gondola. A large spherical descent module (or lander) stood at the apex, balanced on a narrow-waisted orbital module, which in turn stood on an instrument housing. Project engineers mounted its two large

solar panels (each with a span of almost 22 feet—6.7 meters) on either side of the orbital module. The entire spacecraft weighed a hulking 10,800 pounds (4,936 kilograms) at launch and measured 18.7 feet (5.7 meters) long. The powerful Proton rocket served as its launch vehicle.

During the missions of Venera 9 to 12 (all of which adhered to the same basic design), the orbiter separated from the lander as the spacecraft approached Venus. As the orbiter began to circle the planet, the descent module fell through its atmosphere, during which time one parachute opened, followed by another designed to jolt off the upper half of the lander's round shell. Then a third and fourth parachute unfurled, jettisoning the lower half of the sphere, and lowering the exposed instrument package to the surface. The lander contained eight instruments, designed to scan the topography (with a panoramic imaging system), to detect radioactive elements in the soil (with a multichannel gamma spectrometer), and to measure g forces during the entry into the atmosphere (with accelerometers). Meanwhile, as the orbiter revolved around Venus, it activated its instrument suite, which included an imaging system, an ultraviolet detector to sample the clouds, and ion/electron sensors.

Venera 9 lifted off on June 8 and its descent module touched down on Venus on October 22, 1975. First, a black and white camera took panoramic pictures. Then, for fifty-three minutes, data streamed up to the orbiter and back to Earth, revealing a temperature at ground zero of 725 F (485 C) and a pressure of 90 atmospheres. The succeeding spacecraft—Venera 10—made landfall on October 25, 1975, after being launched on June 14, just six days after Venera 9. Its photographs proved to be sharper than Venera 9's and revealed a totally flat desert with boulders scattered in the distance. A density sensor determined that the rock on which the descent module came to rest consisted of properties similar to basalt on Earth. The transmission ended after sixty-five minutes. Veneras 11 and 12, launched on September 9 and 14, 1978, settled onto the landscape on December 25 and December 21, respectively. Neither one returned photos (because their cameras' lens caps failed to release), but both made detailed atmospheric observations, which confirmed that carbonic acid and nitrogen constituted the main ingredients, and that the Venutian clouds consisted of droplets of sulfuric acid.

The last of the Veneras—15 and 16—arrived at Venus in October 1983. Orbiters only, they studied the face of the planet for eight months, mapping 25 percent of the surface. By this point, the intense competition that inspired both the Venus and Mars explorations by the two superpowers began to soften; in mid-1985 Soviet and American geologists joined forces to interpret the images sent home by Venera 16.

Venera 9 (4V-1 No. 660) Probe
Prelanding Configuration

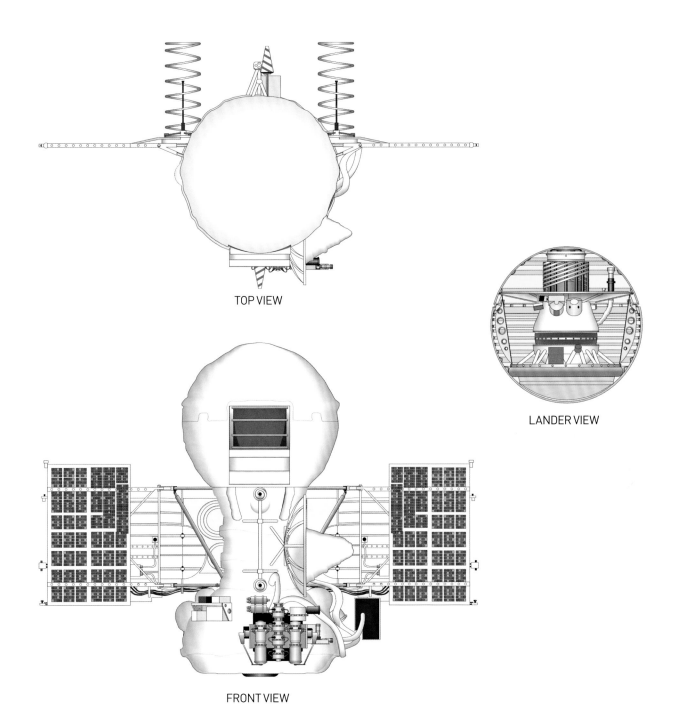

TOP VIEW

LANDER VIEW

FRONT VIEW

0   1   2
                         meters

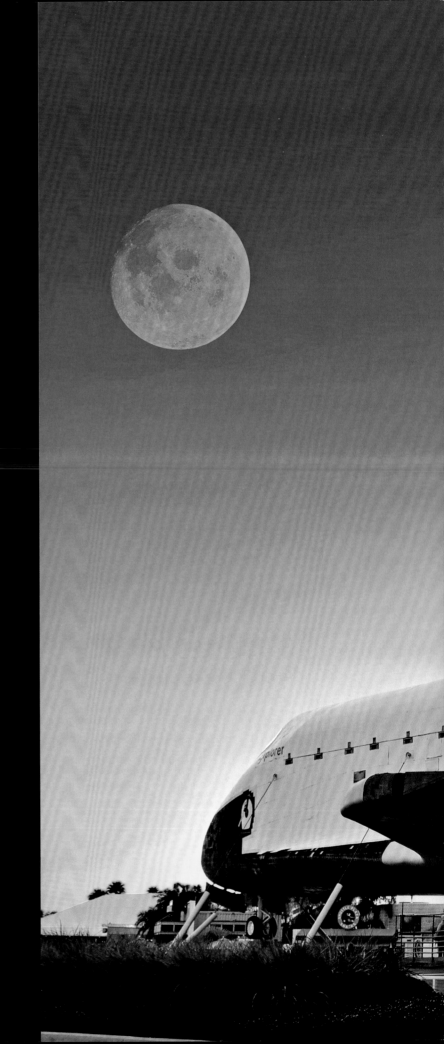

2

THE
SECOND
SPACE
AGE

1977–1997

Soyuz 7K-ST (Soyuz T)

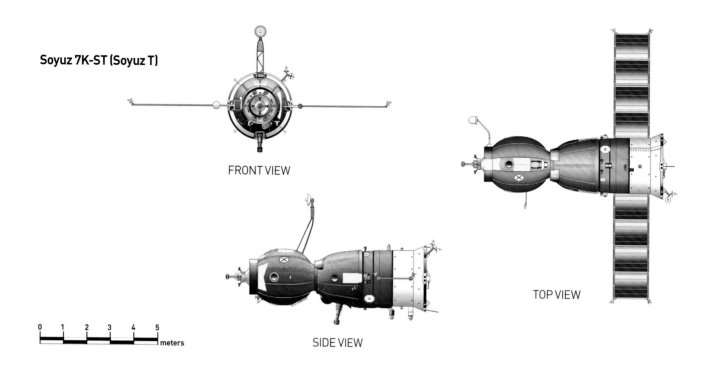

FRONT VIEW

SIDE VIEW

TOP VIEW

0 1 2 3 4 5 meters

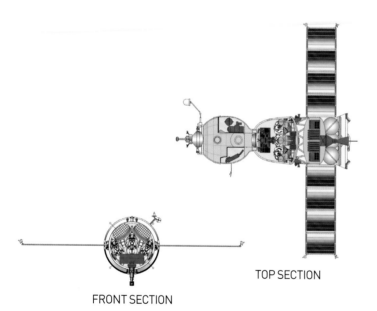

FRONT SECTION

TOP SECTION

SIDE SECTION

0 1 2 3 4 5 meters

CAPSULES
Soyuz 7K-ST

The long line of Soyuz spacecraft continued from the 1960s and 1970s into the 1980s with the 7K-ST, a vehicle that followed in the Soviet tradition of measured, incremental improvements rather than sudden or revolutionary designs. It originated, like the other Soyuz vehicles, with Special Design Bureau 1, founded by Sergei Korolev. During the late 1960s—after Korolev's passing—his organization proposed a military space station, along with a two-seat crew ferry called the 7K-S (the "S" denoting Special, a designator for military projects). Based on the standard Soyuz 7K-OK, it included significant improvements, such as an internal docking system that enabled the passage of cosmonauts *inside* of the Soyuz. The old Soyuz system relied on spacewalks for crew transfers.

Plans called not only for long and short duration variants of the 7K-S (long for space station transport missions; short for special solo missions), but also for a cargo vehicle known as the 7K-G. However, after two years of development (1968–1970), the 7K-S became vulnerable when the military lost interest in its

Soyuz 7K-STM (Soyuz TM)

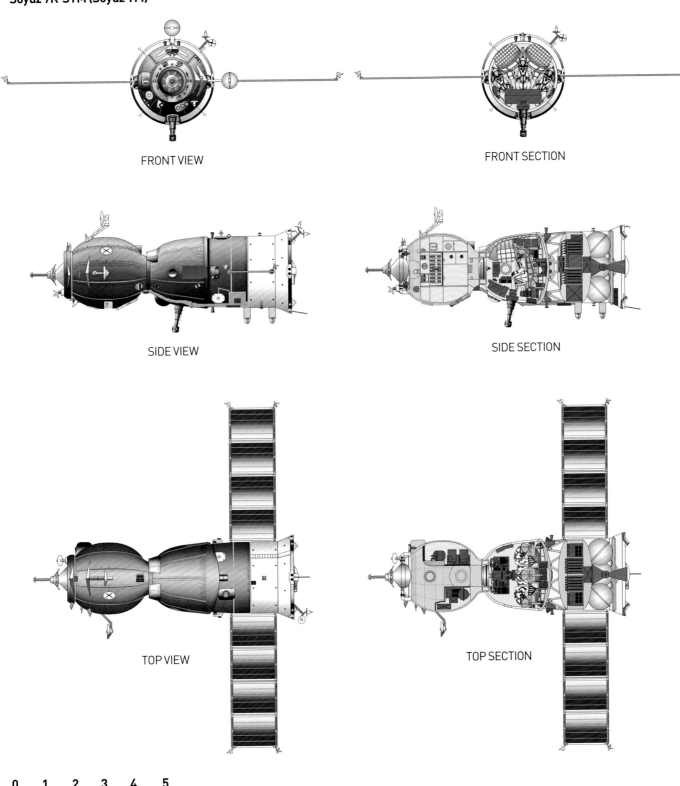

FRONT VIEW

FRONT SECTION

SIDE VIEW

SIDE SECTION

TOP VIEW

TOP SECTION

0 1 2 3 4 5 meters

TKS Spacecraft

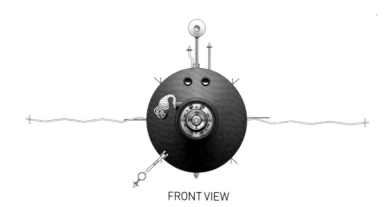

FRONT VIEW

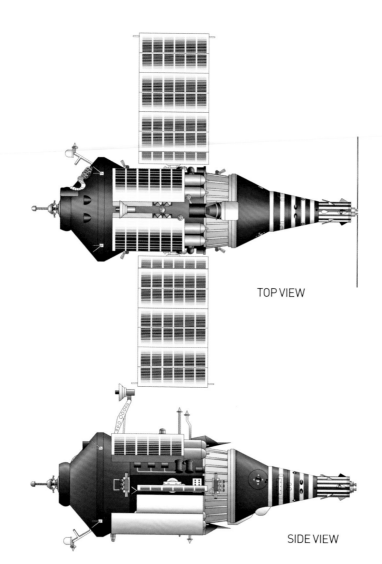

TOP VIEW

SIDE VIEW

0 1 2 3 4 5

meters

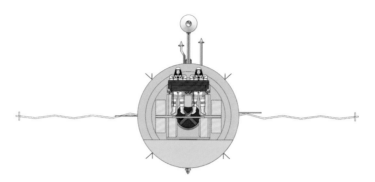

FRONT SECTION

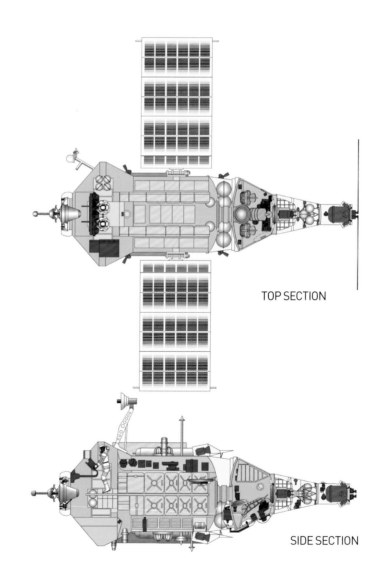

TOP SECTION

SIDE SECTION

space station project. The Soviet armed forces finally abandoned the 7K-S in 1974. But because of its anticipated advantages over the Soyuz 7K-OK and the 7K-T, the 7K-S lived on as engineers at the Korolev design bureau continued to refine it. They eliminated its capacity for independent flight, added new solar panels, and made other changes.

The modified 7K-S finally won acceptance for civilian purposes in 1975, and three unmanned prototypes underwent flight testing from 1974 to 1976. Satisfied with its performance, the Soviet space agency redesignated it the Soyuz 7K-ST and added it to its active roster. A big, heavy spacecraft, the 7K-ST weighed 15,100 pounds (6,850 kilograms) fueled, and measured 24.54 feet (7.48 meters) long, with a span of 34.7 feet (10.6 meters). Despite the passage of time since the first Soyuz in the mid-1960s, its tri-partite design (habitation module forward, descent module in the middle, service module aft) resembled the earliest models.

As a ferry for space station flights, the 7K-ST became a mainstay of the Soviet program, from its first mission in April 1978 (designated Cosmos 1001) until its last in March 1986. It could accommodate three cosmonauts and became the first manned Soviet spacecraft equipped with onboard digital computers. Launched eighteen times, it failed to reach orbit just once and carried cosmonauts to the Salyut 6 and 7 space stations in fifteen of its missions. During these flights, the Soviet space agency often referred to the 7K-ST simply as the Soyuz-T (T for Transport). The Soyuz-T later underwent modifications as the Soyuz-TM, which serviced the Mir and the International Space Station.

Soyuz-TM (7K-STM)

Pressured by the tough Cold War rhetoric of newly elected President Ronald Reagan—which his administration reinforced with the Strategic Defensive Initiative in 1983 and the US space station go-ahead the following year—Soviet political leaders directed the space agency to draft plans for a successor to the Salyut space stations. The resulting Mir went into orbit in 1986.

To complement Mir, engineers at the RKK Energia design bureau considered how to ferry crews to and from the new domicile in space. They answered, in characteristic Soviet space agency fashion, with a solid, incremental solution based on more than twenty years of experience; that is, they improved, but did not fundamentally alter the Soyuz formula. Mir's transport vehicle retained a clear structural kinship with the Soyuz 7K-OK of the 1960s: a three-part spacecraft that consisted of a spherical orbital module forward, a bell-shaped descent module in the middle, and a cylindrical service module aft. Using the most recent ferry (the Soyuz 7K-ST) as a model, in 1981 the Energia engineers proposed improvements and modernizations including upgraded

avionics, better communications, a more durable metal for the body, lighter heat shield materials, a new combined engine unit, and the installation of the Kurs rendezvous and docking system that saved weight and enabled the new spacecraft to maneuver independently of the station. Pounds shed due to the Kurs system also enabled bigger payloads.

The bulky spacecraft that resulted, known as the Soyuz-TM (also called the Soyuz 7K-STM), weighed 15,980 pounds (7,250 kilograms) fueled and measured 24.54 feet (7.48 meters) in length with a span of 34.7 feet (10.60 meters). It made just one preliminary flight: on May 23, 1986, an unmanned Soyuz-TM docked with Mir and undocked six days later, during which time it underwent tests both in independent flight and as a constituent of the station. The Soyuz-TM undertook its first operational mission in February 1987 and its last in April 2002. During its more than fifteen years of service, it flew a total of thirty-four times to Salyut 7, Mir, and the International Space Station. On mission 34, Soyuz-TM—modified as Soyuz-TMA—did not leave the station, but instead remained there, docked to Zarya's nadir port, where it remains in the twenty-first century as a lifeboat on the ISS.

TKS

Time and again, Soviet spacecraft designers adapted their creations incrementally to changing circumstances. The TKS (a Russian designation based on "Transport Supply Spacecraft") epitomized this preference. The TKS originated in 1965 in Chief Designer Vladimir Chelomei's design bureau (OKB-52), a main rival to Special Design Bureau 1. Earlier in the decade, Chelomei's team developed the famed Proton rocket, but their LK-1 manned lunar orbiter lost out to Korolev's Soyuz 7K-L1. After Korolev's death in 1966, Chelomei had better success, winning approval for the TKS.

Rather than a lunar orbiter like LK-1, the TKS served as a ferry to the Almaz military space station. But in contrast to the Soyuz or even the Apollo capsules, the TKS offered an ingenious design with two spacecraft in one. During launch, it consisted of a forward section, the VA (a Russian acronym for "return apparatus"), that looked like an elongated cousin of the Apollo capsule, for transporting crews during launch and reentry; and, linked by a short tunnel, a cylindrical functional cargo block (FGB) at the aft end. The VA came equipped with its own reaction controls, de-orbit braking engine, parachutes, and soft-landing engine. The FGB's large pressurized storage compartment provided resupply for the station, and it also transported docking hardware and maneuvering engines for the combined TKS. In a fitting tribute to adaptability, the VA's engineers derived its design from Chelomei's defunct LK-1.

TKS Family Development

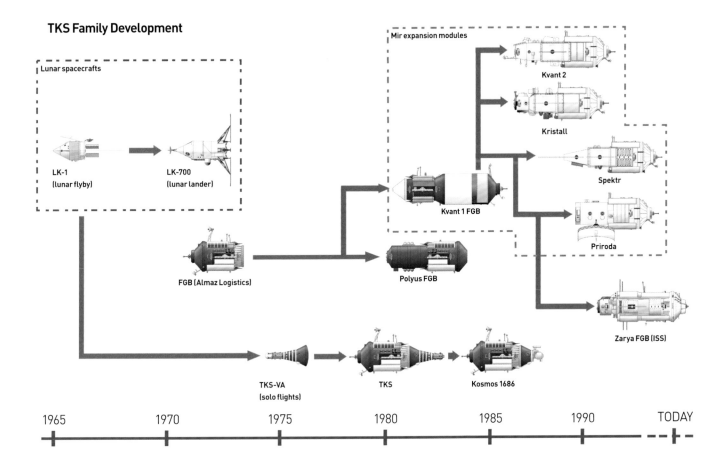

Lunar spacecrafts

LK-1 (lunar flyby)

LK-700 (lunar lander)

FGB (Almaz Logistics)

Polyus FGB

Kvant 1 FGB

Mir expansion modules

Kvant 2

Kristall

Spektr

Priroda

Zarya FGB (ISS)

TKS-VA (solo flights)

TKS

Kosmos 1686

1965   1970   1975   1980   1985   1990   TODAY

Each of the mated halves of TKS could operate independently of the other, with a crew of three, or autonomously (with no passengers at all). The VA could function by itself for up to thirty-one hours and the FGB could fly as an un-crewed cargo module. Together, as a unit, the TKS looked imposing. It measured 43.3 feet (13.2 meters) long with a 13.6-foot (4.15-meter) diameter; its solar arrays spanned 55 feet (17 meters). The TKS offered 1,589 cubic feet (45 cubic meters) of habitable volume, weighed 38,600 pounds (17,510 kilograms), and could carry a payload of 27,700 pounds (12,600 kilograms).

Despite its promise, TKS only underwent four test flights, and no cosmonaut ever manned it. These trials spread out over eight years. On the first mission, launched in July 1977 and designated Kosmos 929, the TKS made a solo test flight with no attempt to rendezvous with a space station. Then, in April 1981, Kosmos mission 1267 sent the TKS to Salyut 6, where the FGB docked, prior to which the VA separated and de-orbited. (On this mission, the FGB proved its space worthiness when it completed fifty-seven days of autonomous flight after it left the Salyut). The first linkup of the complete TKS with a Soviet station occurred in

March 1983 when, in Kosmos 1443, it joined with Salyut 7, after which the VA flew autonomously for four days. Finally, lifting off in September 1985 as Kosmos 1686, the TKS docked with Salyut 7, but this time with a specific purpose—stripped of seats and manned controls, the VA came to the station loaded with instruments: an Ozon spectrometer, a high-resolution photo apparatus, and an infrared telescope.

Despite its limited flying career, the TKS lived on in many other guises; the Soviet talent for adapting designs offered no better example than the TKS. The Mir space station incorporated its technologies on its Kvant-1, Kvant-2, Kristall, Spektr, and Priroda modules, and the ISS's Zarya and Multipurpose Laboratory also descended from TKS.

SPACEPLANES
Space Transportation System

NASA's space transportation system (STS)—like the International Space Station and the Hubble Space Telescope—underwent a long and complicated gestation before becoming reality. Discussions about the STS began in the early 1960s, when a group of engineers in industry and government considered a new way to access space without massive, expendable launch vehicles such as the Saturn V. Anticipating smaller budgets once Apollo ended, they sought alternatives to the extravagantly expensive spaceflight of the time. As these individuals thought about it, an analogous situation existed in the airline industry: just as United and American made initial investments in aircraft and flew them for many years with regular maintenance and repair, reusable rocketry could offer the same advantages in cost and reliability. This concept led to proposals for spaceplanes rather than capsules, and for the refurbishment of rockets, rather than their disposal.

Technical debates over the practicalities of such systems went on throughout the 1960s. At a conference in 1968 hosted by George Mueller, NASA's associate administrator for manned space flight, two main designs emerged. A team from Lockheed outlined an architecture in which, after liftoff, two propellant tanks fell into the ocean for later recovery after boosting a recyclable main module into orbit. Engineers from General Dynamics, on the other hand, recommended mounting a piloted, reusable aircraft similar to the X-15 onto an Atlas missile for the ride into space. Neither these, nor any of the other suggestions, satisfied Mueller. He did, though, refer to some of these prototypes as "space shuttles," capable of transporting astronauts and supplies to a space station then under consideration.

In addition to these concepts sketched out on paper, three actual aerospace research programs sponsored by NASA and the air force presented potential ways to cut overhead and lower the cost of reaching orbit through recycled components. The first test case involved the X-15 aircraft. Flown from 1958 to 1968, its 199 experimental missions proved that piloted vehicles could fly into space; maneuver there; withstand the intense heat of reentry; make unpowered, accurate landings on runways; and be flown again after renovation.

The air force actively pursued a spaceplane of its own known as Dynamic Soaring—shortened to Dyna-Soar—between October 1957 and December 1962, for which it recruited NASA pilots as future astronauts. Dyna-Soar's designers envisioned a one-seat, flat-bottomed, hypersonic glide vehicle powered into orbit by a specially-built two-stage booster. Although the Department of Defense cancelled Dyna-Soar, it spent over $400 million on the project, which yielded profuse data. It also provided concrete planning for a reusable system.

A category of aircraft known as the lifting bodies constituted the third experimental avenue. These small, gum drop- and stiletto-shaped vehicles underwent a series of trials at NASA's Flight Research Center at Edwards Air Force Base, California. Three basic prototype designs flown from 1962 to 1973 proved the feasibility of controlled, runway landings by vehicles returning from space.

In the end, a figure well known inside of NASA catalyzed the debate about a new means of access to space. Maxime A. Faget—the director of engineering and development at the Manned (later Johnson) Space Center in Houston—gained a reputation for forceful advocacy of simple and conservative spacecraft design. Faget, who conceived of the Mercury capsule, entered the shuttle discussions with this formula in mind. He proposed a two-stage configuration consisting of a spaceplane the size of a 707 airliner, launched from an enormous, winged rocket, about as big as a 747 jumbo jet. At the edge of space, booster and aircraft uncoupled, with the rocket falling back to Earth for refurbishment and the spacecraft later making a glide landing much like that of the X-15. Engineers at the Air Force Flight Dynamics Laboratory made one crucial revision to Faget's plan: they argued that a delta-wing shape, rather than Faget's straight-wing, offered greater lift at hypersonic speeds.

Ultimately, the shuttle's proponents found themselves in the political arena, and, surprisingly, they benefited from the encounter. Things looked dire at first when, in 1970, the Nixon administration cancelled a planned space station project, for which the shuttle would serve as a resupply vehicle. Without the station, what would become of the shuttle? NASA acted quickly to shore up its position. Its acting Administrator George Low enticed the air force to participate in the heavy-lift potential of the shuttle by offering to equip it with a vast, 60-by-15-foot (18.3-by-4.6-meter) cargo bay to accommodate military launches. This step helped NASA gain congressional approval for the project. But James Fletcher, the incoming NASA administrator, wanted more; he sought the blessing of the White House. President Nixon gave it, in part because in his upcoming reelection campaign it enabled him to promise high-paying shuttle construction jobs in a number of key states. But he also acknowledged the argument of Caspar Weinberger, his director of the office of management and budget: that cancelling the shuttle, in addition to the space station, threatened all that the United States had gained technically, politically, and psychologically from Mercury, Gemini, and Apollo.

Fletcher got the go-ahead in January 1972 for what became known as the space transportation system. Four firms responded to the request for proposals for the prime contract: North American Rockwell, McDonnell Douglas, Lockheed, and

Columbia (OV – 102)

0 1 2 3 4 5 meters

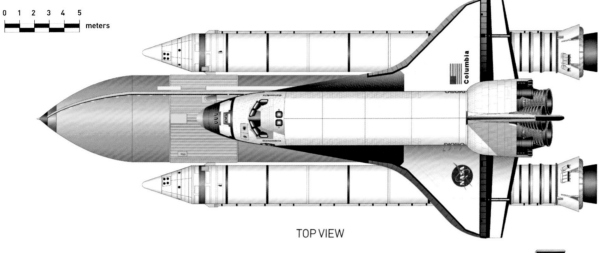

TOP VIEW

SIDE VIEW

FRONT VIEW

Grumman. North American won the $2.6 billion competition, for which it agreed to fabricate two space-worthy orbiters and one full-scale test vehicle. The subcontractors also profited handsomely from the project. Grumman built the orbiter's wings; Rocketdyne its main engines; Thiokol the two solid rocket boosters; and Martin Marietta the massive, liquid propellant external tank. McDonnell Douglas provided overall support. In all, STS startup costs (in 1971 dollars) totaled about $6.75 billion. The NASA Johnson Space Center managed the STS.

Planning for and fabricating the STS had been one thing. Seeing it fueled and ready on launch pad 39A at the Kennedy Space Center on April 12, 1981—the day of the first space shuttle flight—surprised even those intimately acquainted with its evolution. Gigantic in scale, it looked like a machine from a fable, with the two solid rocket boosters and the taller external tank

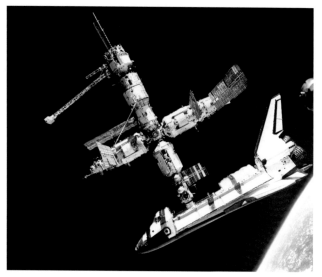

pointing skyward and a full-scale airliner mounted incongruously on the side of the structure. It inspired even greater appreciation as its engines roared and shook, generating over 6.4 million pounds of thrust as this 4.5 million-pound (2.25 million-kilogram), 154-foot (47-meter) tall, 78-foot (24-meter) wide edifice embarked on its maiden voyage.

The Shuttle Orbiters

The shuttle orbiters represented a technological leap over all spacecraft that preceded them, and their operational record stands alone among the world's space vehicles. Although modified during their period of service from 1981 to 2011, they remained fundamentally the same for thirty years—comparable to a DC-9 airliner at 122 feet (37.2 meters) long, with a 78-foot (23.8-meter) wingspan and a weight of about 250,000 pounds at launch. The orbiters flew 135 missions at an average rate of 4.5 per year. They sent 833 crewmembers into space for an aggregate 1,323 days aloft. They hauled more than 3.5 million pounds into orbit, consisting of 180 satellites and other payloads; they returned fifty-two spacecraft and space station components to Earth, weighing more than 229,000 pounds in all. They docked thirty-seven times at the International Space Station and spent 234 days constructing it. They went on seven missions to capture and repair orbiting spacecraft. They remain modern marvels of technology.

But despite all the positive attributes, two horrific shuttle accidents occurred, taking fourteen lives and destroying 40 percent of the shuttle orbiter fleet. And although billed as the answer to high-cost access to space, the total STS expenditures reached approximately $209 billion, or $1.5 billion per mission.

During its tenure, the space shuttle brought America some of its best moments, as well as some of its worst. In the end, Congress approved the construction of five operational orbiters, adding three to the two originally ordered from North American Rockwell. The orbiter saga began with *Columbia*, like the rest of the shuttle fleet named for ships prominent in maritime exploration—in this case, the venerable US Navy frigate that between 1838 and 1840 became one of the first American vessels to circumnavigate the world. It also commemorated the Apollo 11 command module. *Challenger* recalled the Royal Navy research ship that explored the Pacific and the Atlantic Oceans from 1872 to 1876. NASA chose *Discovery* for the third orbiter in memory of the vessels captained by two English adventurers: Henry Hudson, during his search for the Northwest and Northeast Passages in 1610 and 1611, and Captain James Cook, during his famous journey to the Hawaiian Islands and to Western Canada in the 1770s. The fourth orbiter's namesake, *Atlantis*, had more recent

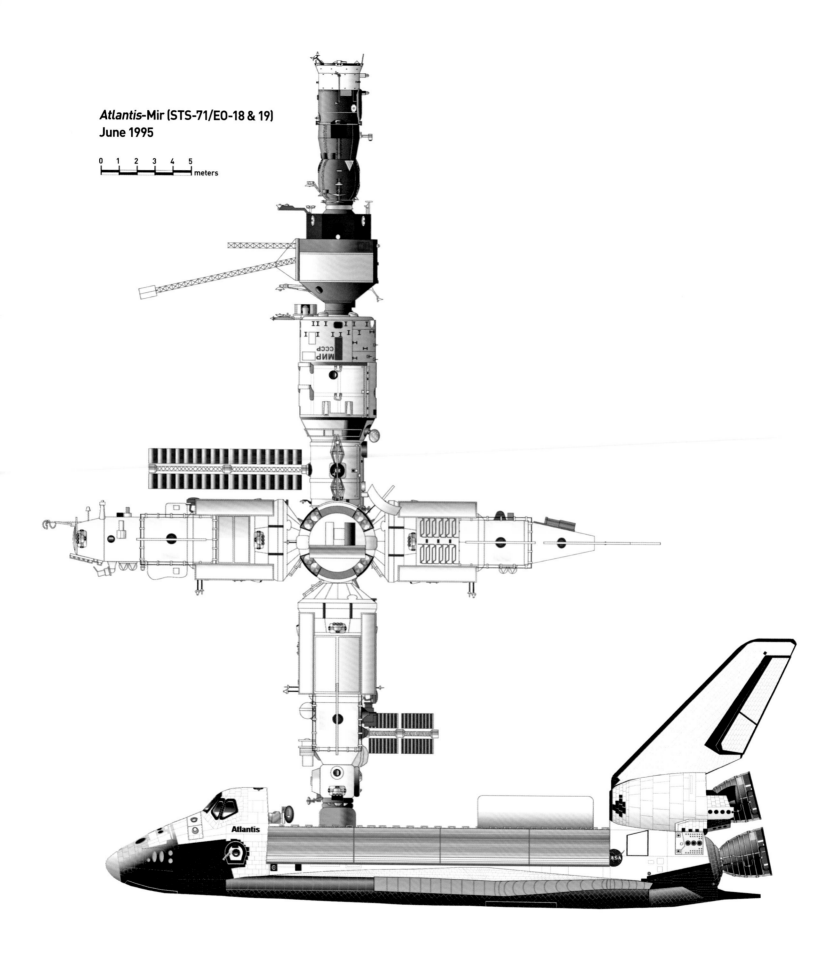

Atlantis-Mir (STS-71/EO-18 & 19)
June 1995

0 1 2 3 4 5 meters

STS Family Development

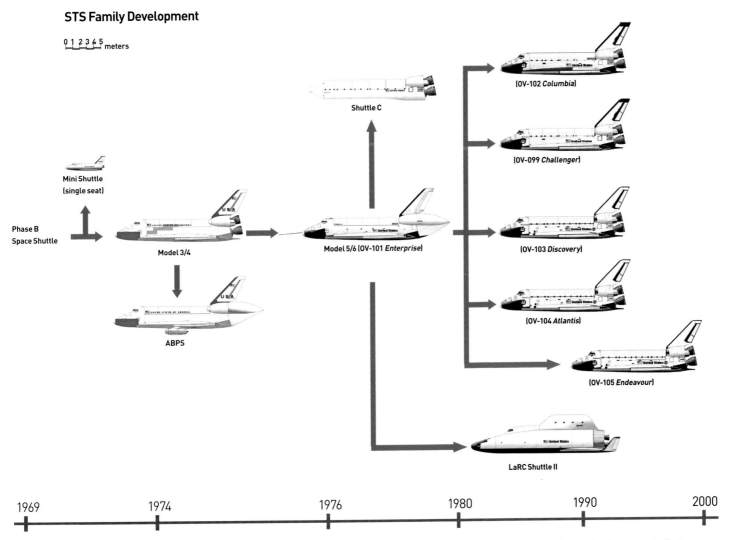

0 1 2 3 4 5 meters

Mini Shuttle (single seat)

Phase B Space Shuttle

Model 3/4

ABPS

Shuttle C

Model 5/6 (OV-101 *Enterprise*)

(OV-102 *Columbia*)

(OV-099 *Challenger*)

(OV-103 *Discovery*)

(OV-104 *Atlantis*)

(OV-105 *Endeavour*)

LaRC Shuttle II

1969   1974   1976   1980   1990   2000

and humble origins, as a two-mast ketch that sailed the oceans for the Woods Hole Oceanographic Institute from the 1930s to the 1960s. Finally, the space agency chose the name *Endeavour* to honor Captain Cook's first command, in which he traveled to Australia and New Zealand from 1769 to 1771. It also recognized the command module of Apollo 15.

North American Rockwell completed the assembly of *Columbia* in March 1979. It represented the heaviest of all the orbiters, with increased versatility over the others. To add to its utility, it underwent modifications that allowed longer missions of nine to twelve days, rather than the usual five to seven. Two years after its rollout, *Columbia* and the massive shuttle stack waited on Pad 39A at the Kennedy Space Center for the launch countdown. It aroused intense public interest, if for no other reason than its appearance. Unlike the familiar, elongated rockets of Apollo days and earlier, the shuttle structure looked like an

immense cathedral, with an aircraft attached unexpectedly to the main vertical component. The liftoff also got the undivided attention of the media and other onlookers, as their bodies resonated with the horrific blast of the solid rocket boosters and the shuttle main engines as they came to life.

Commander John Young and pilot Robert Crippen served as the first crew, taking *Columbia* on a two-day, thirty-six-orbit shakedown voyage beginning on April 12, 1981. It performed well; instrumentation on the spacecraft recorded normal temperatures, pressures, and accelerations on its exterior, and checks of the computer system and cargo bay doors all proved to be satisfactory. But after its landing at the NASA Dryden Flight Research Center on Edwards Air Force Base (the touchdown point for ten of the first twelve shuttle flights), technicians found some unsettling pieces of evidence. They discovered that sixteen of the heat-resistant ceramic tiles that protected *Columbia* from the scorching temperatures of reentry had been lost and 148 damaged. And the trend continued. *Columbia* returned from space in flight two (November 1981) with twelve damaged tiles, and in the third mission (March 1982), with thirty-six missing and nineteen damaged.

Still, the first flights of *Columbia*, *Challenger* (in April 1983), *Discovery* (in August 1984), and *Atlantis* (in October 1985)—indeed, the first twenty-four missions—seemed to belie these or other worries. For instance, during this period, the orbiters enjoyed success in carrying four spacelabs—small laboratories anchored in the cargo hold that conducted science experiments pertinent to physics, astronomy, materials, life sciences, and Earth sensing—for the European Space Agency. These initial forays also began the process of democratizing spaceflight. Because the shuttle accommodated more astronauts than ever before—by January 1986 it had lifted 125 individuals into orbit—it opened participation to Guy Bluford, the first African American, and to Dr. Sally K. Ride, the first American woman.

Even at this early date, however, some of the operational flaws of the shuttle became evident. Clearly, the STS proved to be far less economical than its proponents predicted. And it required months—not days or weeks, as once hoped—for technicians to refurbish previously flown parts for reuse. The process demanded big and costly engineering and maintenance crews.

But on January 28, 1986, disillusionment gave way to despair. On that unusually cold Florida morning, ground inspections of launch pad 39A at Kennedy Space Center revealed significant accumulations of ice. After a two-hour postponement, *Challenger*—flying STS-33, the twenty-fifth shuttle mission—lifted off. A faint puff of smoke emanated from the field joint located on the lower portion of the right solid rocket booster. Eight more puffs occurred in the first 2.5 seconds, and at 59 seconds flame appeared. Six seconds later, the plume broke through the wall of the external tank. At 72 or 73 seconds after launch, an explosion of immense power resulted in a massive failure of the entire structure. Seven astronauts died in the catastrophe: Commander Richard Scobee; Pilot Michael Smith; Mission Specialists Dr. Judith Resnik, Dr. Ronald McNair, and Col. (posthumous) Ellison Onizuka; electrical engineer Gregory Jarvis; and teacher Sharon Christa McAuliffe. President Ronald Reagan led a stunned nation in mourning.

An independent commission headed by former Secretary of State William P. Rogers determined that the disaster happened when prolonged cold temperatures caused the rubber O-rings on the right SRB to shrink, resulting in a leak and finally, in flames being exhausted onto the external tank. The Rogers Commission also faulted Thiokol, the SRB's manufacturer, for failing to warn

NASA that the cold temperatures on January 28 posed a threat to a safe launch.

As a result of the *Challenger* accident, no shuttle flights occurred for two years and eight months as the space agency tried to correct deficiencies. Under the supervision of James Fletcher, who returned to NASA as administrator for a second time during this crisis, the SRBs underwent a thorough redesign, and to fill the gap in the fleet left by *Challenger*'s loss, the agency contracted with North American to build the shuttle *Endeavour*. With the flight of STS-26 in September 1988 aboard *Discovery*, the shuttle returned to operation.

During the remainder of its service life, not one flight or flights can claim to be the most significant. But two sets of multiple missions stand out as lasting legacies of the space shuttle era. During the nineteen years from 1990 to 2009, NASA devoted six flights to launching, repairing, and updating the famed Hubble Space Telescope. These daring events in 1990, 1993, 1997, 1999, 2002, and 2009 represented an immense advancement in long-duration extra vehicular activities (EVAs) and proved the practicality of complex, lengthy on-orbit repairs. They also transformed Hubble from an embarrassment—a costly telescope blinded by astigmatism—to an instrument that produced mesmerizing images of the universe in the visible and near infrared spectra, with a probable life expectancy of about thirty years.

The other outstanding accomplishment of the orbiters involved the construction of the massive International Space Station. Achieved over thirteen years, from 1998 to 2011, three orbiters (*Atlantis*, *Endeavour*, and *Discovery*) carried most of the laboratories, control modules, solar arrays, airlocks, trusses, girders, and the other paraphernalia necessary to erect this great structure. The ISS required thirty-one assembly missions; of that number, twenty-seven flew on the orbiters.

The beginning of the end of the space shuttle program happened unexpectedly, on the morning of February 1, 2003. *Columbia* went into orbit as STS-107 on January 16, 2003, with a crew of seven who (among other things) tested the physiology of space endurance, observed climate over central Africa and elsewhere, and conducted agricultural experiments. Not surprisingly, its members consisted of relatively young, highly accomplished men and women: Commander (Col.) Rick Husband and Pilot (Cmdr.) William McCool; Mission Specialists (and medical doctors) Capt. David Brown and Capt. Laurel Clark; Mission Specialists Dr. Kalpana Chawla and Lieut. Col. Michael Anderson; and Col. Ilan Ramon of the Israel Air Force. During their reentry, less than five minutes after they entered the atmosphere, a sensor on the left wing's leading edge spar

Columbia (OV-102)

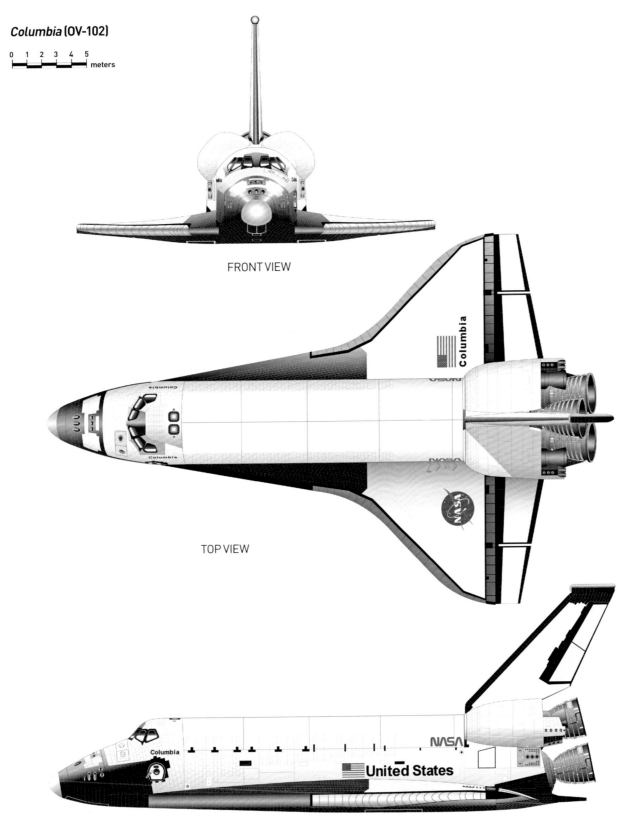

0 1 2 3 4 5
meters

FRONT VIEW

TOP VIEW

SIDE VIEW

Columbia (OV-102) and the Shuttle Orbiters

0 1 2 3 4 5 meters

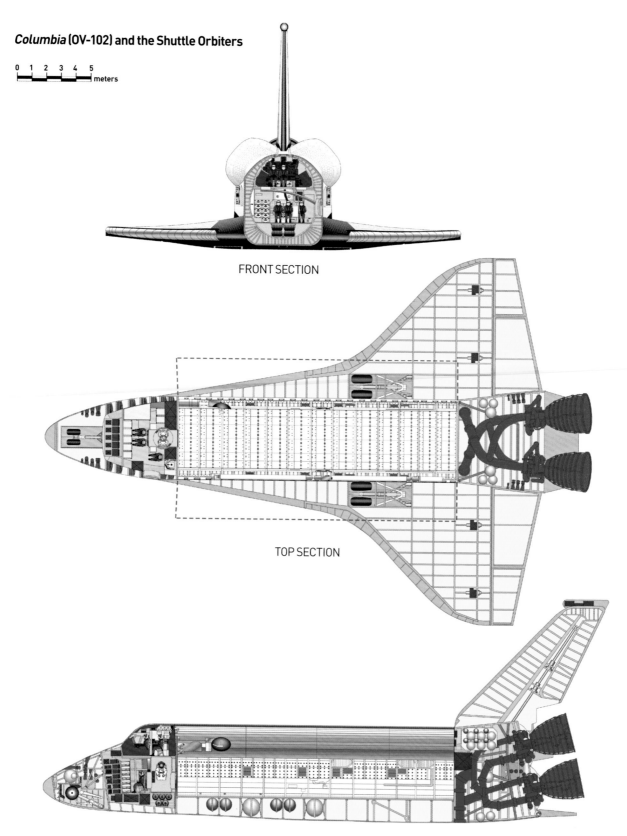

FRONT SECTION

TOP SECTION

SIDE SECTION

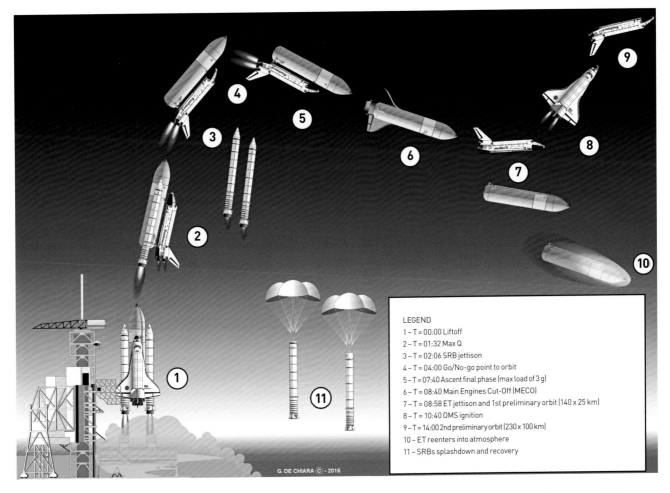

LEGEND
1 – T = 00:00 Liftoff
2 – T = 01:32 Max Q
3 – T = 02:06 SRB jettison
4 – T = 04:00 Go/No-go point to orbit
5 – T = 07:40 Ascent final phase (max load of 3 g)
6 – T = 08:40 Main Engines Cut-Off (MECO)
7 – T = 08:58 ET jettison and 1st preliminary orbit (140 x 25 km)
8 – T = 10:40 OMS ignition
9 – T = 14:00 2nd preliminary orbit (230 x 100 km)
10 – ET reenters into atmosphere
11 – SRBs splashdown and recovery

G. DE CHIARA Ⓒ - 2016

showed higher strain than normal. As the orbiter streaked across the night sky over California, Nevada, Utah, Arizona, and New Mexico, and onward toward Texas, onlookers on the ground saw debris trailing from the arc of light. Mission control in Houston, meanwhile, saw that four sensors in the left wing failed. Pieces of *Columbia* rained down as it traveled eastward into Texas from the New Mexico border. NASA called an emergency and named retired Adm. Harold Gehman to chair the Columbia Accident Investigation Board (CAIB).

After six months, in August 2003 the CAIB issued its findings: the accident occurred because a piece of insulating foam on the external tank broke loose and struck the Orbiter's left wing, breaching some of its thermal tiles. On reentry, the exposed area heated and melted the underlying aluminum structure of the wing, causing its failure, and finally that of the entire orbiter. But the Gehman group also expressed an essential truth about the STS: that despite initial aspirations and promises, it never became fully reliable and cost-effective—the commercial airliner of space travel, as its developers hoped—but remained an experimental

vehicle, always subject to risks and failure. The panel further suggested that in light of the dangers inherent in the shuttle and its already long service, the time might come soon for it to be retired. President George W. Bush responded in January 2004 with a decision to terminate the shuttle in 2010.

After the return to flight in July 2005 by *Discovery*, the space shuttles flew twenty-two more missions, all but one of which relating to ISS construction or supply. To accomplish its responsibilities, NASA extended its flight manifest to July 8, 2011, on which date *Atlantis* delivered the Rafaello multipurpose logistics module to the station. With that, the space shuttle era ended.

Orbiter *Enterprise*

If the rocketry for the space transportation system (STS) inaugurated a new age in access to space, the orbiter itself represented one of the most significant engineering feats of the age. The shuttle orbiter—comparable in size to a DC-9 airliner at 122 feet (37.2 meters) long, with a 78-foot (23.8-meter) wingspan—weighed roughly 250,000 pounds (113,440 kilograms)

at launch. Its massive 60-by-15-foot (18.3-by-4.6-meter) cargo bay hauled objects as big as a standard railroad boxcar. Its 50-foot- (15.2-meter-) long remote manipulator arm not only grappled outsized payloads of all shapes and weights, but also grasped very small objects with remarkable dexterity.

Surprisingly, no real technological midpoints existed between the Apollo capsule—an archetype of the previous generation of human space exploration—and the orbiters. Shuttle's designers made the improvements in one leap. They replaced limited pilot options in Apollo with a real cockpit and a full range of controls. They replaced an incredibly cramped instrument panel and seating layout with captain's chairs and

a true flight deck. They eliminated the hazards of splashdown at sea and replaced it with precise runway landings. Perhaps most tellingly, the interior space of Apollo versus the shuttle orbiters reflected the progress that had been made. The Apollo command and service module that took Neil Armstrong, Michael Collins, and Buzz Aldrin to the moon contained just 218 cubic feet (6.2 cubic meters) of habitable area; an orbiter crew (up to seven) lived in 2,525 cubic feet (71.5 cubic meters), spread out over the flight deck, the mid-deck/ equipment bay, and the airlock. In other words, the orbiters gave the astronauts almost twelve times more livable space than America's previous space transport.

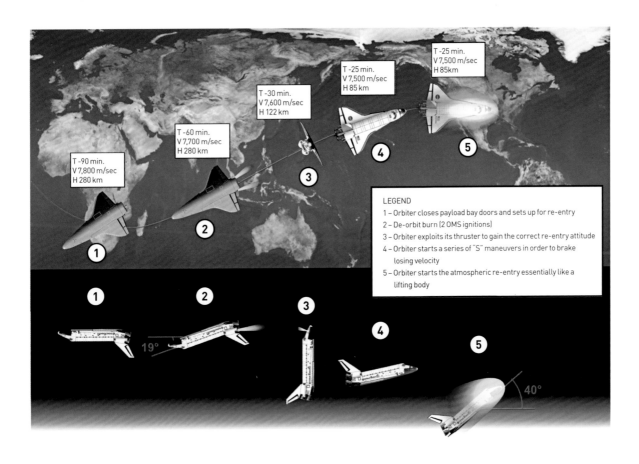

T -90 min.
V 7,800 m/sec
H 280 km

T -60 min.
V 7,700 m/sec
H 280 km

T -30 min.
V 7,600 m/sec
H 122 km

T -25 min.
V 7,500 m/sec
H 85 km

T -25 min.
V 7,500 m/sec
H 85km

LEGEND
1 – Orbiter closes payload bay doors and sets up for re-entry
2 – De-orbit burn (2 OMS ignitions)
3 – Orbiter exploits its thruster to gain the correct re-entry attitude
4 – Orbiter starts a series of "S" maneuvers in order to brake
    losing velocity
5 – Orbiter starts the atmospheric re-entry essentially like a
    lifting body

19°

40°

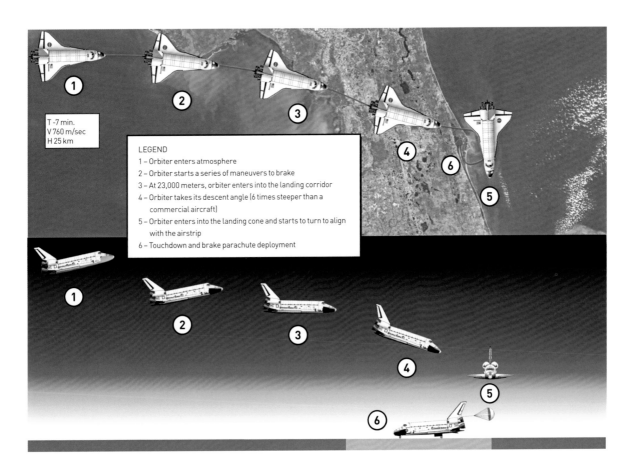

T -7 min.
V 760 m/sec
H 25 km

LEGEND
1 – Orbiter enters atmosphere
2 – Orbiter starts a series of maneuvers to brake
3 – At 23,000 meters, orbiter enters into the landing corridor
4 – Orbiter takes its descent angle (6 times steeper than a
    commercial aircraft)
5 – Orbiter enters into the landing cone and starts to turn to align
    with the airstrip
6 – Touchdown and brake parachute deployment

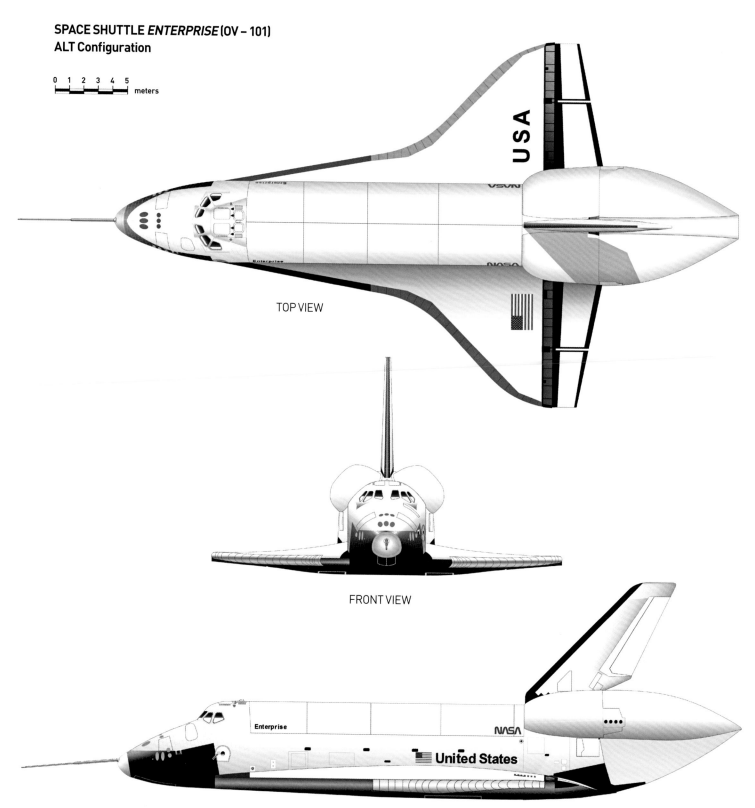

SPACE SHUTTLE *ENTERPRISE* (OV – 101)
ALT Configuration

0 1 2 3 4 5 meters

TOP VIEW

FRONT VIEW

SIDE VIEW

North American Rockwell won the prime contract for the shuttle in 1972. It subcontracted out all components but the orbiters, leaving that key part for its own engineering team. Before construction got underway, NASA asked North American for some design changes based on about forty-six thousand hours of NASA and air force wind tunnel tests. The data suggested several key modifications: reducing the orbiter's original weight (accomplished by making the wings smaller and lighter); reshaping its planform into more blended wing-body contours; and making the heat-resistant tiles that covered the Orbiter's exterior as uniform as possible. At the end of the process, the orbiter measured 3 feet shorter than originally planned and featured a more flowing nose-body section, a smaller nose radius, and a revised airfoil profile.

To assess the orbiter's flying qualities in the atmosphere, NASA ordered a flight research prototype without engines from North American, in addition to the two space-worthy orbiters under fabrication. NASA's leaders wanted to call it *Constitution*, but a vigorous letter-writing campaign to President Gerald R. Ford persuaded the space agency to bow to the more romantic name preferred by the public: *Enterprise*, after the starship in *Star Trek*. It rolled out of the North American plant in Palmdale, California, in September 1976 and at the end of January 1977, the national media watched as it inched along on a flatbed truck to the nearby NASA Dryden Flight Research Center on Edwards Air Force Base.

Under the direction of project manager Deke Slayton at the Johnson Space Center in Houston, Texas, the engineers at Dryden prepared a flight plan that required *Enterprise* to undergo five unpowered missions known as approach and landing tests (ALT). They wanted to ascertain pilot impressions of the orbiter's actual flying qualities, but also to gather hard data about its behavior by mounting extensive instrumentation on the vehicle. The team needed a jumbo carrier aircraft for the project and considered a 747 airliner and a military C-5 cargo aircraft. They decided

on the 747 because of its superior handling characteristics and procured an older model that they used not only for the ALT tests, but also later for the ferry flights that transported orbiters from Edwards to the Kennedy Space Center in Florida. Perched atop its mother ship on three pillars, *Enterprise* and the 747 underwent eight ground and flight tests from February to July 1977 to determine their combined aerodynamic compatibility before the ultimate trial: uncoupling *Enterprise* from the 747 at more than 20,000 feet and allowing it to glide to a runway on Edwards.

These five ALT flights uncovered some important deficiencies in the orbiter's systems. The first took place in August, the next two in September, and a final one in October, with two of them commanded by astronaut Fred Haise of Apollo 13 fame.

Pressure mounted prior to the last *Enterprise* launch because in each of the first four the two-man crew missed the landing targets and reported chattering and ineffectiveness in the brakes. When Haise stepped into the cockpit again for the fifth mission, flown on October 26, 1977, a lot went through his mind. He regretted the four missed targets in the earlier flights and told himself to press hard to land *Enterprise* right at the marker on the runway. He also flew it for the first time without its customary tail cone in place, which changed its aerodynamics. Haise might even have been distracted by the intense media coverage, heightened by the presence of Charles, the Prince of Wales, at the event.

In any case, after *Enterprise* uncoupled from the carrier aircraft and Haise took command, he worked the controls in a tight, compact manner and came in nose-high and a little fast. He used the stick to adjust the orbiter's sink rate, which resulted in swings in *Enterprise*'s elevons, in turn causing pitch up (nose-to-tail oscillations). As he approached the runway, Haise gave commands to the onboard computer to damp these motions, but just as he touched down, the spacecraft jumped up, bouncing off the surface. As it hung in the air during the next four seconds, violent wing-to-wing, pilot-induced oscillations (PIOs) shook

Haise and his co-pilot Gordon Fullerton. *Enterprise* seemed ready to crash, but Fullerton shouted, "Hey, let loose." Haise took his hands off the controls, the PIOs lessened, and they landed safely.

This discovery of pitch-up and PIOs in the orbiter's handling characteristics changed the course of space shuttle development. After extensive simulated approaches aboard YF-12 and Navy F-8 test aircraft, researchers at Dryden realized that on the fifth ALT flight, *Enterprise*'s mechanical control surfaces responded too slowly to Haise's computer inputs, which in part explained the nearly catastrophic outcome. As a remedy, Dryden engineers devised software filters that restrained, but did not entirely suppress PIO and pitch-up tendencies in the orbiters' flight profile.

By 1979, the shuttle team at the Johnson Space Center applied these flight research lessons, incorporating the Dryden modifications to the two orbiters still under construction.

VKK Buran

Buran ("snowstorm" in Russian) got its start in the heated climate of suspicion that characterized the Cold War. Soviet military leaders became convinced that the space shuttle existed not for the reasons claimed by NASA—for scientific research and space physiology—but rather, for the darker purposes of aiming lasers or even nuclear warheads at the USSR. So, to keep pace in the race for weaponry, during 1974 and 1975, the Energia Design Bureau under the direction of Valentin Glushko conceived of a spaceplane at least superficially similar to the shuttle. The Central Committee of the Communist Party approved Buran in 1976. By that time, the space shuttle had advanced far in its development. During the year after Buran's initiation, the shuttle underwent full-scale glide tests to ascertain its handling qualities and aerodynamics. For their part, the Soviets watched these and other events with keen interest. In response to the US Air Force's Dyna-Soar boost-glide military spaceplane and NASA's lifting body research in the 1960s and 1970s, they launched their own lifting body project, named Spiral, in 1966, culminating in the test flights of the subsonic MiG 105-11 between 1976 and 1978. Spiral ended as Buran began.

Once Buran underwent its final design phase, fabrication began in 1980, just as the shuttle prepared to begin routine operations. Six scale-model flights followed in 1983 and a full-sized prototype of Buran became available in 1984, outfitted with four jet engines. Rather than copy the method used to test the shuttle's handling qualities—mounting an orbiter on top of a 747 carrier aircraft and then releasing it, unpowered, on a glide path to the runway—the Buran test model took off from the ground and when it reached its designated altitude, its pilot shut off the engines and flew an approach and landing. Engineers at Energia put Buran thorough a rigorous flight research program of twenty-four missions between 1985 and 1988, each with a crew of two.

The spacecraft and booster that emerged from this period looked so much like the space shuttle that many believed that the Soviets merely imitated the American design. But the obvious similarities disguised important and intentional differences. Those who conceived of Buran looked upon the shuttle as a predecessor and as a first attempt; worth emulating, but also subject to modification. They analyzed each of the shuttle's main features and decided, in most cases, to adopt them because they could not be substantially improved. But in a few key areas they made critical changes, which ultimately resulted in a new generation of rocketry.

Rather than relying on the shuttle's twin solid rocket boosters (the source, after all, of the *Challenger* accident), the Soviet designers used four liquid-propellant rockets (which afterward became the basis of the Zenit launch vehicle). And instead of positioning three main engines on the orbiter itself, like the shuttle, the Soviets installed four main engines on a separate, massive rocket stage. Although this design prevented the restoration and reuse of any of its main components, it also enabled the Buran to haul up to 95 tons of payload in the launch system itself (contrasted with a maximum 29 tons carried in the shuttle's cargo bay). Just as Zenit spun off from Buran, the heavy-lift launch vehicle proposed for the Soviet shuttle found a separate but short life as the Energia rocket.

Buran and the shuttle orbiter did not differ significantly in size or mass. The Soviet vehicle measured 119.3 feet (36.37 meters) in length, with a wingspan of 78.5 feet (23.92 meters) and a payload bay 60.9 feet (18.55 meters) long by 15.3 feet (4.65 meters) in diameter. It had a maximum weight 231,000 pounds (105,000 kilograms). The Energia rocket consisted of four RD-170 liquid oxygen/kerosene engines for the boosters and four RD-0120 liquid oxygen/liquid hydrogen engines for the core stage.

Buran made only one orbital launch. It lifted off on November 15, 1988, from the Baikonur Cosmodrome, Khazakhstan, and, in a fully automated flight with no cosmonauts on board, conducted a flawless mission of two orbits, lasting 206 minutes, before a making precision landing.

Despite this success—exceptional for a maiden voyage—Buran faced extinction. Not only did the end of the Cold War result in the disintegration of the USSR in 1991 and the disruption of the Soviet space agency's activities, but funding diminished and by 1993 Buran vanished entirely from the budget. More than that, it expired in part because many of its detractors felt that it represented a technology chasing a mission—a criticism often leveled at the space shuttle as well.

VKK Buran (OK-1K1)

0 1 2 3 4 5
meters

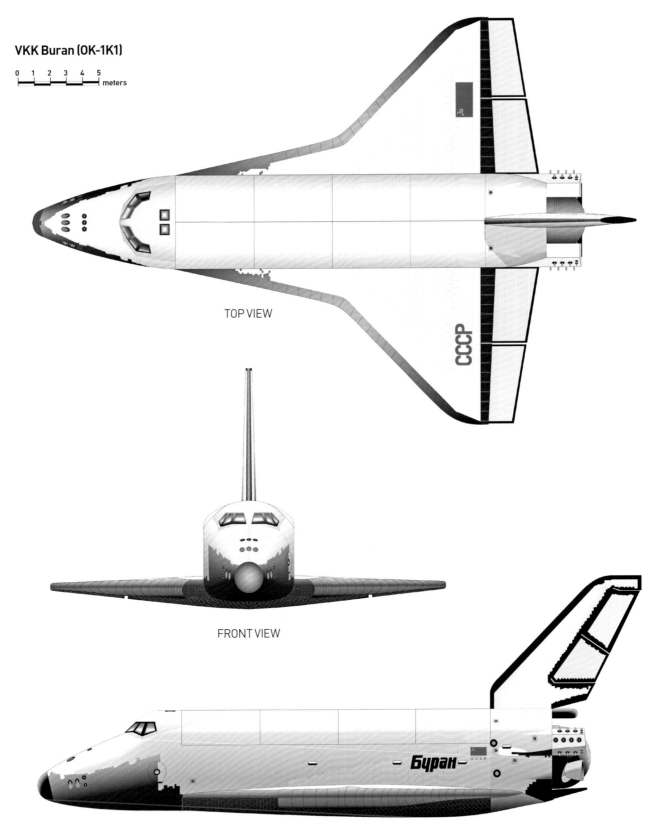

TOP VIEW

FRONT VIEW

SIDE VIEW

Salyut 7 (DOS-7K #6)

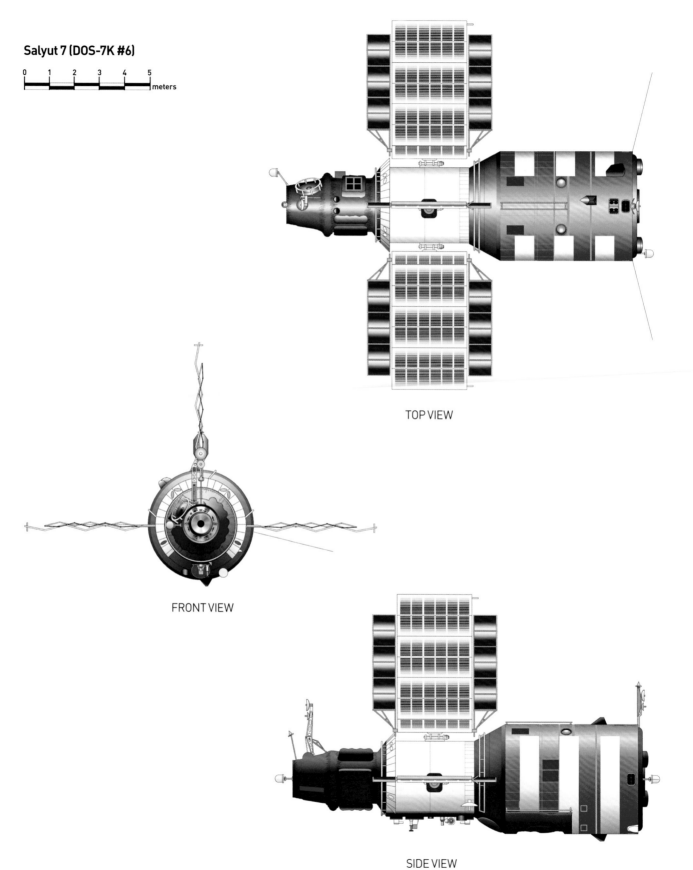

0 1 2 3 4 5 meters

TOP VIEW

FRONT VIEW

SIDE VIEW

STATIONS
Mir

Although the prospect of grand, ringed space stations seized the American imagination during the 1950s, the United States had neither the first nor the longest experience with these habitations on high. That distinction went to the USSR.

Soviet prominence in this area originated with its loss to the US in the race to the moon. From that point in 1969, long-duration spaceflight became what the Soviets pursued for decades. From April 1971 to August 1977, Salyuts 1 and 4 (civilian space stations) and Salyuts 2, 3, and 5 (military stations known covertly as Almaz) sent crews to dock with and inhabit these vehicles for brief visits. Then came a breakthrough in Salyut 6, which, from 1977 to 1982, conquered the challenges of living in space for extended periods.

As Ronald Reagan ramped up the Cold War during his early presidency with such programs as the Strategic Defense Initiative in 1983 and the US space station go-ahead in 1984, Soviet Premier Yuri Andropov approved a successor to Salyut/Almaz. This new station, Mir ("Peace"), went into orbit in February 1986, and although its structure resembled Salyuts 6 and 7, its designers refashioned the interior, turning it into a core habitation vehicle equipped with a five-port docking unit. Even at its start, Mir weighed over 23 tons and measured 43.6 feet (13.3 meters) long.

From this base, it expanded as new modules arrived: the Kvant 1 ("Quantum"), added in April 1987, initiated astronomical observations with a laboratory bay and transfer compartments; Kvant 2, which arrived in December 1989, facilitated EVA activities and improved the quality of habitation (through the addition of water regeneration units, oxygen production hydrolysis equipment, a shower, and an airlock); the Kristall module, installed in June 1990, provided materials processing facilities (furnaces and biotechnology units), as well as docking ports for eventual shuttle visits; the Spektr extension, launched in May 1995, carried remote sensing instruments (both Russian and American) related mainly to materials science; and Priroda, which reached the station in April 1996, enabled the remote sensing of Earth's atmosphere, its land, and its oceans, to which a consortium of European nations contributed.

By far the biggest spacecraft up to its time, its modules increased the overall size of Mir to 62.3 feet (19 meters) long, 101.7 feet (31 meters) wide, and 90.2 feet (27.5 meters) tall. It weighted a majestic 285,940 pounds (129,700 kilograms).

During its fifteen-year lifespan, Mir hosted crews of two to four cosmonauts, who received provisions from the automated Progress resupply vehicle every two to three months. It broke all records for long-duration spaceflight, most notably the one set by Dr. Valeri Polyakov who stayed on Mir continuously for 437 days. In all, it domiciled twenty-eight long-duration crews (as distinct from visitors who stayed aboard for about a week, during handovers from one crew to the next), and it welcomed 104 different individuals from twelve countries.

Like most space vehicles, Mir experienced problems—in its case, ones proportionate to its size. Two collisions and a fire blemished its service record. One crash happened in January 1994, when a departing crew in a Soyuz TM-17 hit the Kristall module at least twice as it conducted an inspection of the station before deorbiting. Before the next impact, the unthinkable occurred: a fire broke out in February 1997 during a routine ignition of an oxygen canister. The crew put on gas masks and extinguished the flames, although the cabins and the Soyuz lifeboat filled with smoke (cleared eventually by the air filtration system). The second collision took place in June 1997, when a remote-control docking test caused the cargo vehicle undergoing maneuvers to tumble and strike the Spektr module. The crew heard the hiss of escaping air but managed to cut the cables feeding the affected area, stopping the leak.

Despite these harrowing events, the experiences of Mir's designers, engineers, cosmonauts, and flight operations team contributed heavily and directly to the success of the International Space Station.

Atlantis-Mir

When President Ronald Reagan announced in his State of the Union Address in January 1984 that his administration planned to embark on a massive space station project, he did so knowing that the USSR, the other Cold War superpower, had a much

OPS/DOS Family Development

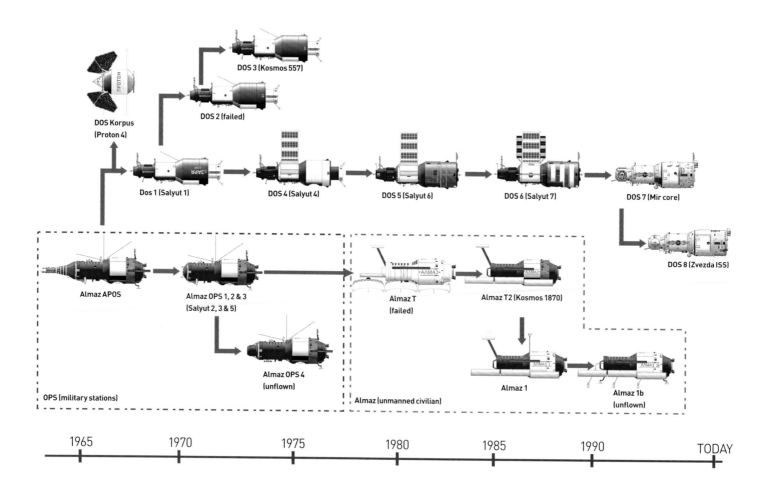

0 ––––– 5 meters

DOS 3 (Kosmos 557)

DOS Korpus (Proton 4)

DOS 2 (failed)

Dos 1 (Salyut 1)

DOS 4 (Salyut 4)

DOS 5 (Salyut 6)

DOS 6 (Salyut 7)

DOS 7 (Mir core)

DOS 8 (Zvezda ISS)

Almaz APOS

Almaz OPS 1, 2 & 3 (Salyut 2, 3 & 5)

Almaz T (failed)

Almaz T2 (Kosmos 1870)

Almaz OPS 4 (unflown)

Almaz 1

Almaz 1b (unflown)

OPS (military stations)

Almaz (unmanned civilian)

1965   1970   1975   1980   1985   1990   TODAY

longer history with long-duration spaceflight than the United States. The extent of NASA's experience occurred in the four Skylab missions in 1973 (that combined for 171 days in space). The Soviets, on the other hand, began work on Salyut ("Salute") 1 during the 1960s and launched it in April 1971. But American eagerness to pursue a station did not stem only from an attempt to cut Russia's lead; it originated, in part, from the knowledge that Salyuts 2, 3, and 5 actually served as cover for three secret, military spacecraft, named Almaz ("Diamond") 1, 2, and 3. Taken as a whole, the seven stations flown under the Salyut banner (including Salyut 6 and 7) gave the Soviets fifteen years of trial and experimentation, until the series ended in 1986.

The Soviet space agency supplanted Soyuz in 1986 with the Mir station, launched in February 1986 on a Proton rocket. Meanwhile, President Reagan's plan for an American station

began to encounter stiff congressional opposition that only accelerated with time. Then, in 1993, NASA Administrator Daniel Goldin, constrained by tight budgets, broke with tradition and proposed a radical approach to his Russian counterparts: to combine the talents of both countries in a multinational space station endeavor. A model for joint effort already existed in the 1975 Apollo-Soyuz Test Project (ASTP), which offered valuable technical and organizational precedents for merging the two sides. As part of a broader agreement, the Russian and American space agencies committed themselves to a series of shuttle-Mir rendezvous missions that enabled astronauts to experience long-term habitations on the Russian station, while cosmonauts familiarized themselves with shuttle operations and technology.

Although the shuttle-Mir project produced plenty of favorable publicity, it also represented an open-air laboratory in which the

Mir (DOS-75 #7)

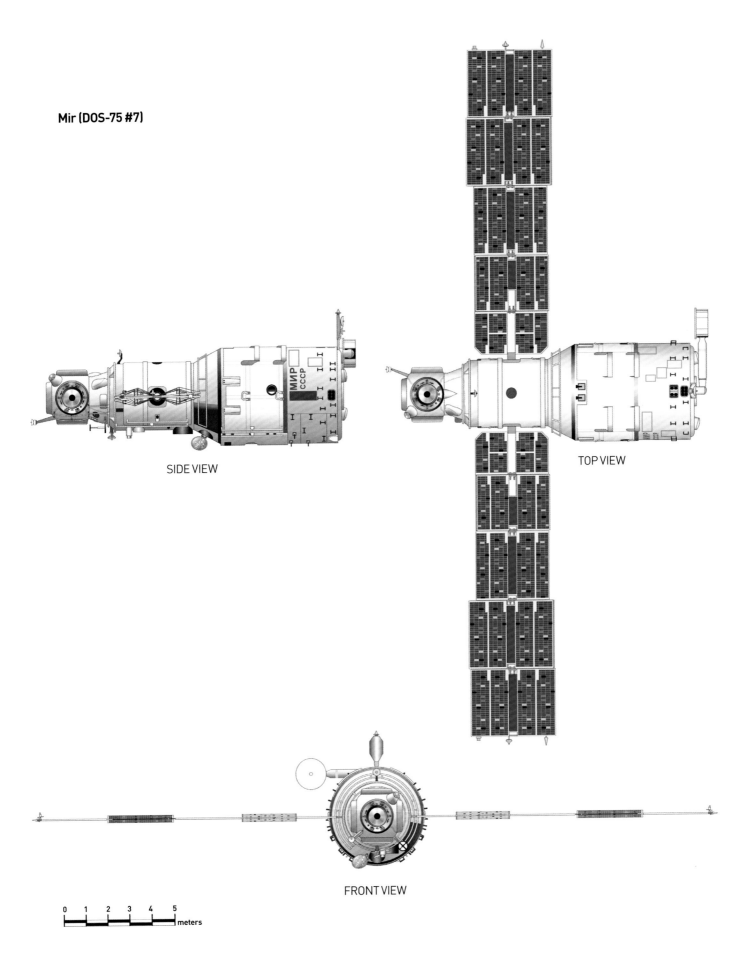

SIDE VIEW

TOP VIEW

FRONT VIEW

0 1 2 3 4 5
meters

two main space powers could transcend their competitive urges and cultural differences to act in concert. The raw numbers bear out its serious purpose: the program lasted for four years and four months, involved seven astronauts and seven cosmonauts, and ended with a total of eleven visits by the space shuttle to Mir.

Two preliminary flights began with the shuttle *Discovery*. From February 3 to 11, 1994, STS-60 carried Sergei Krikalev, the first cosmonaut to fly aboard the shuttle. The next mixed-crew flight (STS-63) happened from February 3 to 10, 1995, in which the astronauts hosted Vladimir Titov, a veteran of the Soviet program. During this mission—a warmup to upcoming dockings—*Discovery* approached Mir to within 40 feet (12.2 meters), then retreated to about 400 feet (122 meters) and flew around the station.

America's initial presence on Mir proved to be a special occasion for several reasons. Astronaut Norman Thagard—the first American to train in Russia and to be launched into space on a Russian rocket (the Soyuz TM-21)—also became the first American on Mir. By the time he left the Russian station (on STS-71), he held the longest endurance record of any American astronaut: 115 days (March 14 to July 7, 1995).

Thagard returned home on *Atlantis*—the only shuttle that ever docked with the Russian station. The second linkup with *Atlantis* (on STS-74) occurred from November 12 to 20, 1995, followed by STS-76 from March 22 to 31, 1996, in which *Atlantis* ferried astronaut Shannon Lucid to her destination. She remained there for 188 days, the longest of any American who flew on *Atlantis-Mir*. Her visit initiated two uninterrupted years of US presence on the station.

After Lucid's time ended in September 1996, *Atlantis* carried five more astronauts from that date to June 1998: John Blaha, Jerry Linenger, Michael Foale, David Wolf, and finally Andy Thomas, who closed the program.

During their long stays on Mir, the American astronauts got more than just a taste of daily, pedestrian life. During Linenger's time on board (January to May1997), he experienced the most severe fire ever encountered by any spacecraft, as well as an uncontrolled tumble when the crew lost attitude control. On Foale's mission (May to October 1997), a Progress supply vehicle collided with Mir, pushing the station into a spin and temporarily causing depressurization.

In the end, when *Atlantis* pulled back from Mir for the last time on June 8, 1998, a new chapter in space history dawned. By then, seven American astronauts spent a total of over 907 days in orbit. During these extended habitations, both American and Russian space programs experienced a greater appreciation for each other's machines and technologies, but they also resolved crises together, expanded their understanding of the physiological effects of protracted periods in space, and, perhaps most significantly, increased their comprehension of the cultural differences that distinguished their two countries. These lessons did not wait long to be applied; just six months after the close of Atlantis-Mir, Russia and the United States docked Zarya to Unity, the first two components of the International Space Station.

Salyut 6 and 7

Salyuts 6 and 7 represented the payoff of a gamble wagered by the Soviets in the early 1970s: to rebuild its reputation after losing the race to the moon by leading the world in extended endurance in space. The bet took a while to be redeemed; the first five Salyuts (the civilian Salyuts 1 and 4; the Almaz military stations that flew undercover as Salyuts 2, 3, and 5) showed promise, but in their six years of service (February 1971 to February 1977), they only achieved marginal success as long-duration spacecraft. The breakthrough came with Salyut 6, which launched on September 29, 1977, and reentered the atmosphere on July 29, 1982. Its success came not because of a radical transformation of the Salyut design, but from steady, pragmatic improvements. In fact, compared to Salyut 4, Salyut 6 offered few structural differences. Both built by Sergei Korolev's Special Design Bureau 1 and carried out by his successor Vasily Mishin, the main modification involved the addition of a second docking port, borrowed from the Almaz configuration conceived by the design bureau of Vladimir Chelomei. The original one had been at the forward end of the spacecraft; Mishin's engineers added the new one at the aft end. This one change influenced station operations profoundly. For the first time, the Soyuz capsule could remain docked—or return with fresh crews—while the Progress freighter arrived with supplies and fuel or hauled away refuse. It enabled permanent habitation.

Salyut 6 also distinguished itself from Salyuts 1 and 4 in having a dual-chamber powerplant, derived from the Almaz stations. Called the unified propulsion system, it allowed the station's twin engines and control thrusters to run on the same propellant: unsymmetrical dimethylhydrazine and nitrogen tetroxide. Additionally, Salyut 6 carried a new telescope, the BST-1M multispectral that made observations in submillimeter, infrared, and ultraviolet light.

The greatest gulf between Salyut 6 and its predecessors involved the number of crewed missions and their time in orbit. In all, sixteen cosmonaut teams (of two or three each) visited the station during its lifespan (September 1977 to July 1982). Of that number, a quarter achieved long-duration occupancy: 96 days for

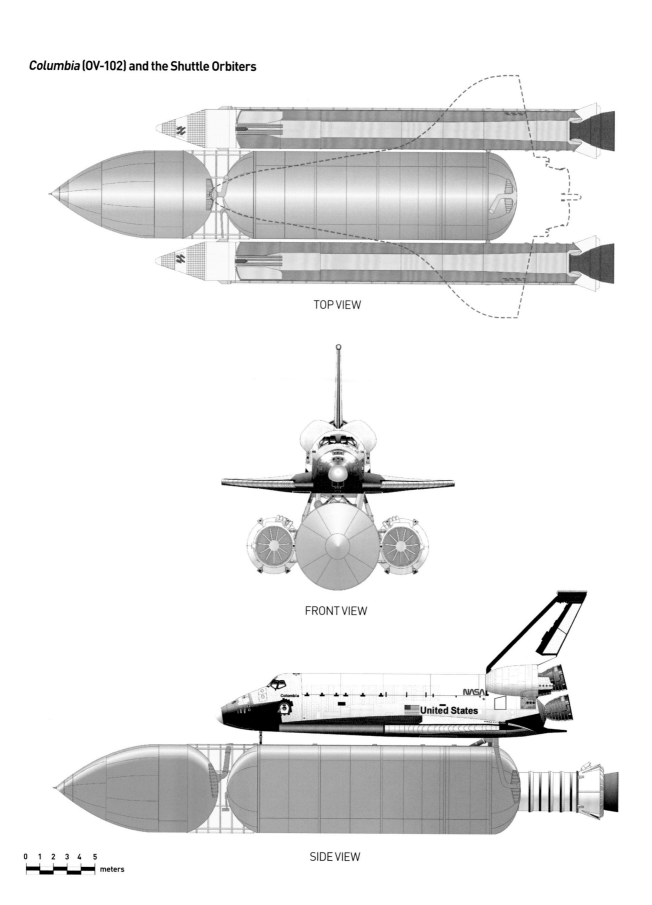

Columbia (OV-102) and the Shuttle Orbiters

TOP VIEW

FRONT VIEW

SIDE VIEW

0 1 2 3 4 5 meters

the first mission, almost 145 days for the fourth, and 175 days for the seventh. But on the eighth mission (Soyuz 35)—led by Commander Leonid Popov and Flight Engineer Valery Ryumin—the crew stayed aloft for six months, from April 9 to October 11, 1980, a few hours short of 185 days. With the completion of this mission, they set a world endurance record.

Salyut 7 (the last of the Salyut line) went into orbit on April 19, 1982. Intended originally as a backup to Salyut 6, it differed only negligibly from its predecessor, although technicians removed the BST-1M telescope and replaced it with an x-ray detection system. Salyut 7 carried crews from May 1982 to July 1986.

Twelve cosmonaut groups, of one to three members, entered Salyut 7, and half of the missions lasted one hundred days or longer. On February 8, 1984, Commander Leonid Kizim, Flight Engineer Vladimir Solovyov, and cosmonaut-physician Oleg Atkov arrived at the station and remained until October 2, 1984, breaking the old longevity mark with nearly 237 days in orbit; that is, more than seven months of continuous habitation. Although certainly not free of serious problems—on one occasion, a fuel leak required four EVAs to fix; on another, all systems on the station went dead, leaving Salyut 7 tumbling in orbit—in the end the Salyut series proved the practicality of residing in space.

And with the launch of the Mir station in February 1986, the Soviet space agency justifiably claimed preeminence in maintaining life outside of the home planet for extended periods.

ROCKETS
Shuttle Rocketry

Because the space shuttle embodied a spacecraft, an aircraft, and a launch vehicle all in one, it had no real predecessors. Among its many groundbreaking attributes, its rocket system may have been the most revolutionary. Unlike the Saturn V, its famous but more traditional predecessor, the shuttle divided its lifting power among three distinct sources. It also introduced the concept of partial reuse, by which the orbiters made runway landings and the solid rocket boosters (SRBs) went back to the Kennedy Space Center after retrieval from the ocean. Both parts then underwent refurbishment and returned to service. Only the shuttle external tank required replacement after each flight. Once it separated from the orbiter at the end of the boost phase (about 70 miles above the Earth), it fell through the atmosphere, and those pieces that did not burn up sank to the bottom of the sea.

During shuttle liftoffs, the twin SRBs came to life first in the launch sequence. At minus 6.6 seconds in the countdown, they awakened and burned quickly, providing 71.4 percent of the total thrust of the shuttle. They cut off at roughly 130 seconds and separated at about 28 miles (45 kilometers) altitude, after which they made a ballistic descent lasting roughly four minutes. Prior to impact into the ocean, three parachutes opened on each SRB. Recovery ships typically found them about 140 miles (225 kilometers) downrange. Onlookers could not help being impressed by the power and profile of the SRB. At 149 feet (45.46 meters) tall with a mass of 1.3 million pounds (589,690 kilograms), the solid propellants alone (ammonium perchlorate, aluminum, iron oxide, a polymer binder, and an epoxy curing agent) weighed 1.1 million pounds (503,950 kilograms). The two boosters provided a combined 5.3 million pounds of thrust.

Unlike the SRBs that gave a quick burst of power to lift the STS off the launch pad and on its way, the three space shuttle main engines (SSMEs) embedded in the rear of the orbiter offered sustained thrust for 8.5 minutes—the duration of the shuttle's powered flight. This slow, long burn accelerated the shuttle from 3,000 miles per hour (4,828 kilometers per hour) to 17,000 miles per hour (27,358 kilometers per hour) during the six minutes required to reach orbit after jettisoning the SRBs. Its propellant consisted of cryogenic liquid hydrogen and liquid oxygen, at a ratio of 6:1, respectively. In all, the three SSMEs added 1,125,000 million pounds of thrust to the shuttle launches.

Finally, the shuttle's rocket system depended on an immense external tank (154 feet/47 meters high, 25.5 feet/8.5 meters in diameter) that not only served as the liquid fuel receptacle for the SSME powerplant, but also as the structural backbone for the entire edifice, absorbing the total thrust loads of the two solid rocket boosters and the three shuttle main engines. The interior consisted of three compartments: one holding the liquid oxygen in the forward position; one holding liquid hydrogen in the aft; and a collar-shaped inter-tank connecting the two. Because of a big disparity in their weights, the portion holding the lighter hydrogen measured 2.5 times that of the oxygen. A coating 1-inch (2.5 centimeters) thick of spray-on polyisocyanurate foam covered the outside of the entire external tank, keeping the propellants at acceptable temperatures while protecting against aerodynamic heating and ice formation.

And yet, despite the historic innovations in the shuttle's rocketry system, it proved to be much less than its designers expected. Although early proponents of the shuttle assumed that reusable rocketry represented the key to affordable orbital access, it remained prohibitively expensive: NASA calculated that during the first twenty missions alone, Americans paid $257 million in direct costs, rising to $286 million by 1990. Many of these advocates also believed that the shuttle held the prospect of frequent, predictable flights, like

Ariane 44-L
(V-150 16 April 2002)

0 1 2 3 4 5
meters

a commercial airliner. But technical complexity and frequent maintenance and repair not only disrupted shuttle service, but also contributed to higher expenditures.

The most disappointing facet of the space shuttle involved its two fatal accidents. Fourteen total deaths occurred among the crews of *Challenger* in 1986 and *Columbia* in 2003, which shattered the reputation of the STS and contributed heavily to its demise in 2011.

Ariane 4

During 1958—the year in which the United States responded to the Sputnik challenge with the launch of Explorer 1 and the birth of NASA—an Italian and a French scientist proposed that the European governments establish an institution devoted to space science, modeled on the European Organization for Nuclear Research (CERN). Two years later, representatives from ten European countries formed a commission to study collaboration in space exploration. Acting on the commission's report, the European powers created two space agencies in 1964, separating rocketry from spacecraft development: the European Launch Development Organisation (ELDO), and the European Space Research Organisation (ESRO). These two entities merged to form the European Space Agency (ESA) in 1975, uniting the efforts of eleven countries: Belgium, Denmark, Germany, France, Ireland, Italy, Netherlands, Spain, Sweden, Switzerland, and the United Kingdom. Canada became a cooperating state in 1978.

Anticipating the needs of ESA, the member countries authorized development of a common rocket (Ariane 1) in July 1973, with management under the supervision of the French space agency, known as the National Center for Space Studies (Centre Nationale d'Etudes Spatiales, or CNES). The CNES joined forces with France's Aerospatiale, a state-owned manufacturer, to design and build this initial launch vehicle. Thirty-six aerospace companies, thirteen banks, and CNES combined in 1980 to form Arianespace, a satellite launch and marketing company to which ESA transferred its new rockets once it completed testing and fabrication.

Ariane 1—a three-stage, liquid propellant booster 154 feet (47 meters) long that carried up to 4,080 pounds (1,850 kilograms) of payload—flew for the first time on December 24, 1979. It made ten more launches (eight successfully) until ESA retired it in 1986. Ariane 2 and Ariane 3 shared the same basic structure as Ariane 1, but with distinctions. Only Ariane 3, which became operational in 1984, had two solid propellant strap-on boosters, and, like Ariane 2, measured 6.5 feet (2 meters) longer than Ariane 1. Both carried heavier payloads than Ariane 1: 4,552 pounds (2,065 kilograms) for Ariane 2; 5,688 pounds (2,580 kilograms) for Ariane 3.

While Ariane 4 relied largely on the technology, hardware, and the experience of Arianes 1, 2, and 3, its designers, in their pursuit of heavier lift, added some features that enabled it to become Europe's workhorse rocket through the 1990s.

Ariane 4 came under consideration by ESA in 1981 and got the go-ahead in 1982. It flew for the first time in 1988 and represented a significant advance in performance over its sister rockets. Engineers at ESA lengthened the first and third stages, strengthened the structure as a whole, added the Spelda dual launch system for releasing two spacecraft at once, and installed new propulsion bay layouts and avionics. The biggest change came with strap-on boosters that used either solid or liquid propellants interchangeably on five models: the 42L, 44L, 42P, 44P, and 44LP. The base 40 type used no boosters.

Because of these measures, Ariane 4 had much greater capability than the earlier Arianes. More than 192 feet (58.4 meters) tall and 12.4 feet (3.8 meters) in diameter, it hauled up to 16,756 pounds (7,600 kilograms) to low Earth orbit. It weighed (depending on variant) as much as 1,036,175 pounds (470,000 kilograms) gross and produced about 1,400,000 pounds of thrust. Its first stage consisted of four liquid propellant Viking 5 engines fueled by a hydrazine-nitrogen tetraoxide mixture, as well as either two or four strap-on boosters, depending on need. The second stage used the same propellants as the first stage, which powered a single Viking 4 engine. Stage 3 contained one HM7B engine, fed by the cryogenic liquid oxygen and liquid hydrogen.

The Ariane 4 flew from 1988 to 2003. Over its fifteen years of service, it made 117 launches with only three failures: in February 1990, January 1994, and December 1994. From March 1995 to the end of the program in February 2003, it completed seventy-four consecutive successful flights. Its cargoes varied, but it flew many communications satellites into orbit for several countries, such as Saudi Arabia's Arabsat 1C in February 1992, Spain's Hispasat in September 1992, and Brazil's Brasilsat in March 1995. Ariane 4 also lifted vehicles conducting science and defense missions, including the NASA/CNES TOPEX/Poseidon ocean topography spacecraft in August 1992 and the French intelligence-gathering Clementine in December 1999. In all, it accounted for 50 percent of the world's commercial satellite launches during its lifespan.

Newly designed and far more powerful than its predecessor, Ariane 5 made its initial launch in June 1996.

Energia

The demise of the Soviet Union and end of the Cold War represented profoundly disruptive challenges to Russian space ambitions. In its aftermath, the Russian economy suffered as it attempted to transform itself abruptly from a planned to a market system. Buffeted by political and economic turmoil, the government reduced funding for space, and with it the ability to pursue massive projects such as VKK Buran—a spaceplane resembling the space shuttle—and its launch vehicle, the Energia rocket. Energia, like Buran, started long before anyone guessed at the USSR's disintegration and the future space partnership with America. Both spacecraft originated in 1976, at a time of high Cold War tensions during which Soviet military planners became convinced—despite NASA's claims to the contrary—that the planned space shuttle represented a potential delivery system for laser weapons or nuclear warheads. This assumption sent the Soviet space program on a crusade to build its own shuttle, complete with an even more powerful rocket system than its American counterpart.

The initiation of Energia coincided roughly with the cancellation of the mighty N1 rocket, designed to transport cosmonauts to the moon. In effect, it was the Soviet answer to NASA's Saturn V. But the N1 failed miserably. Rushed through testing and impeded by the death in 1966 of Chief Designer Sergei Korolev, it suffered four unsuccessful launches, the second of which caused one of the largest nonnuclear explosions in history. Although at first not thought of as a successor to N1, the Energia rocket eventually earned an expanded role. In addition to serving as a two-stage booster for Buran, its designers also configured it as a stand-alone, three-stage rocket capable of launching a wide variety of payloads into orbit.

Whether lifting Buran or other cargo, Energia did not fall too far short of the N1 in size or power. It measured a giant 318 feet (97 meters) tall, 25.4 feet (7.75 meters) in diameter, and weighed 5,565,700 pounds (2,524,600 kilograms). It relied on eight rocket engines: one RD-170 (liquid oxygen/kerosene) for each of its four boosters and four RD-120s (liquid oxygen and liquid hydrogen) for the core stage.

Energia made its maiden flight on May 15, 1987, carrying the Polyus military spacecraft, with a third stage borrowed from a cancelled Mir module. Its first two stages performed well, but Polyus fell into the ocean when the upper stage failed to launch it into orbit. Energia flew again on November 15, 1988, this time with an unmanned Buran on its back. After two orbits and 206 minutes aloft, it made an automated landing at Baikonur.

Despite this preliminary success, political realities conspired against Energia before it could really begin. The dissolution of the USSR in 1991 and the general economic slide that accompanied it impacted the funding of space activities, and by 1993 both Buran and Energia—perhaps the most ambitious space projects ever undertaken by Russia—disappeared from the budget.

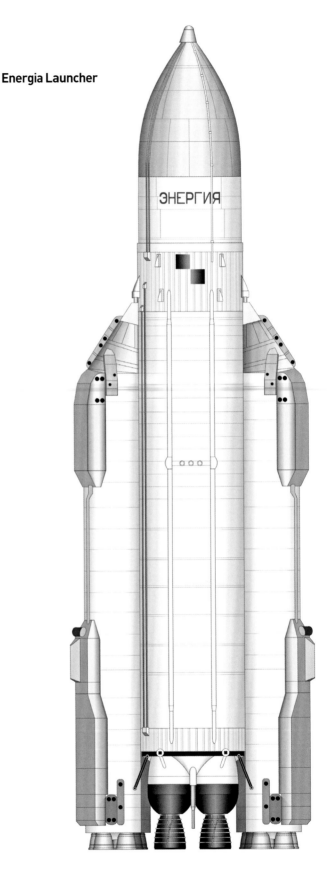

Energia Launcher

0 1 2 3 4 5 meters

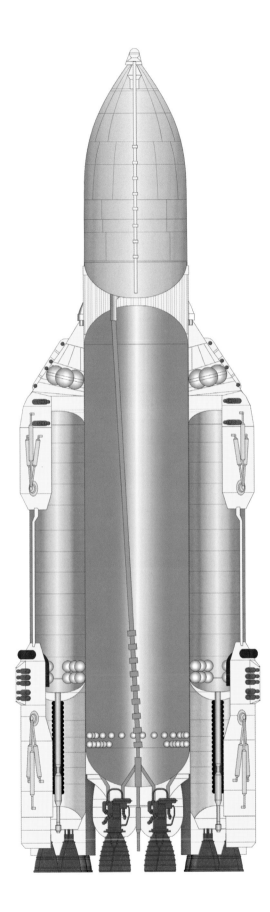

Even so, a glimmer of optimism remained. Engineers at Energia hoped to keep the program alive by advocating for the Energia-M, a smaller rocket that they thought of as a placeholder while they awaited the return of the full-scale Energia. Although weaker than the original (in the core stage, it used two RD-170 engines instead of four, and just one RD-120 instead of four), Energia-M still would have been one of the bigger launch vehicles of its time. But in 1995, it also fell victim to the fiscal axe. Its demise, like Buran's and Energia's, marked a pivot point in the Russian space program. As its leaders faced the realities of Russia's economic slump, they seized on the warmer relations with the United States to enter a new phase. Collaboration—embodied during the 1990s by the two former rivals joining forces to develop the International Space Station—became the new standard.

ROBOTICS
Cassini-Huygens

A joint NASA-European Space Agency (ESA) project to explore Saturn, Cassini-Huygens featured contributions from seventeen countries. An American team at the Jet Propulsion Laboratory (JPL) designed and fabricated the Cassini orbiter spacecraft; the European Space Technology and Research Center managed the Huygens lander. The Italian Space Agency (ASI) provided several of Cassini's science instruments, much of its radio system, and its high-gain antenna.

The name of this probe honors Italian astronomer Jean-Dominique Cassini (1625–1712) and Dutch scientist Christiaan Huygens (1629–1695), who respectively confirmed and discovered the existence of Saturn's rings. Fittingly, the Cassini-Huygens mission concentrated on the Saturnian system, including the planet, its rings, and its eight major moons.

Much of the lore of Cassini-Huygens involved its mighty mass and size: it represented the heaviest object launched into deep space up to that time at 12,593 pounds (5,712 kilograms). It measured 22 feet (6.7 meters) high and 13 feet (4 meters) wide. Despite its proportions, it looked unprepossessing: like a gold-colored, heavy-set insect attached at one end to a large dish antenna. It required the heavy-lift capacity of the Titan IVB-Centaur rocket, America's most powerful expendable launch vehicle, capable of generating 3.4 million pounds of thrust. On the launch pad, the entire stack rose 184 feet (56 meters) from the ground—as tall as a twenty-story building.

Given the complexity of the spacecraft, its mass (like that of an African elephant) and the length of its journey (2 billion miles; 3.2 billion kilometers) rendered this project exceptionally difficult. And these factors fail to account for the combined

Cassini and Huygens Probes

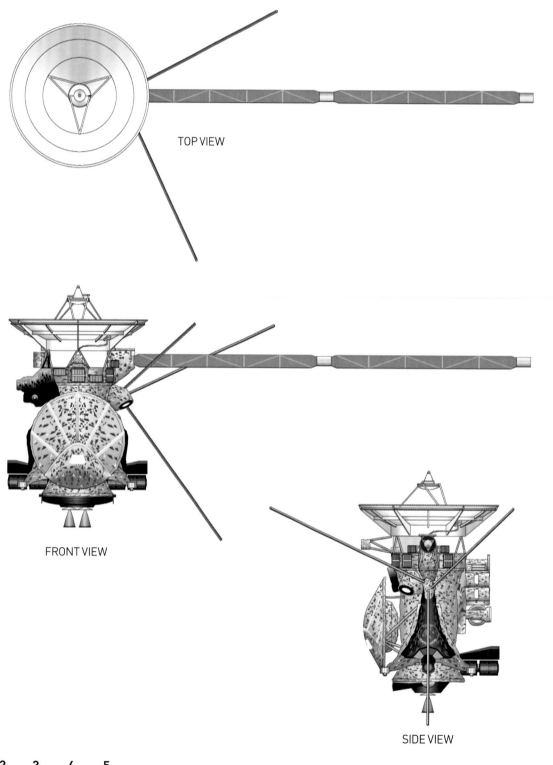

TOP VIEW

FRONT VIEW

SIDE VIEW

0 1 2 3 4 5

meters

Huygens mission to descend into the atmosphere and touch down on the surface of the moon Titan, undertaken while Cassini orbited Saturn.

Cassini-Huygens lifted off in October 15, 1997. It then picked up speed as it flew by Venus in April 1998 and June 1999, passed the Earth in August 1999, reached Jupiter in December 2000, and at last arrived at Saturn in July 2004, at which time its onboard rocket engine fired to brake its speed and drop it into orbit.

As Cassini became oriented to its orbital routine, the team at ESA prepared to launch Huygens on its mission to Titan. A clamshell-shaped spacecraft that measured almost 9 feet (2.7 meters) in diameter and weighed 770 pounds (350 kilograms), its controllers released it on December 25, 2004, for an autonomous, three-week descent at 13,400 miles per hour (21,563 kilometers per hour) to the moon's surface. As the probe sensed Titan's upper atmosphere, three parachutes opened and it drifted for two and a half hours through the foggy, nitrogen–filled atmosphere. During the descent and approach, six instruments on Huygens captured more than one thousand images, sampled Titan's thick atmosphere from top to bottom, measured winds and temperatures, and mapped its surface as it penetrated its haze and clouds. It landed safely on January 14, 2005, and continued to broadcast data to Earth for seventy minutes through the Cassini orbiter.

A tribute to engineering as much as to science, Huygens represented an exclamation point on Cassini's long and fruitful career. The four-year original mission ended in July 2008, a period during which it circled the planet about seventy-five times, flew forty-four flybys of Titan, and concentrated on Saturn, its ring system, its icy moons, and the magnetosphere. Then in the first extension, the Cassini Equinox Mission from July 2008 to October 2010, the spacecraft passed Titan twenty-seven more times and the icy and geologically active moon Enceladus seven times. Finally, the Cassini Solstice Mission, scheduled from October 2010 to September 2017, investigated the seasonal influences on Saturn's rings and satellites, in addition to revisiting Titan on fifty-six more occasions and Enceladus on another twelve. As it made its many passes, Cassini also discovered seven new moons: Methone, Pallene, Polydeuces, Daphnis, Anthe, Aegaeon, and the temporarily named S/2009 S 1.

Cassini's final phase of discovery ended in late summer 2017, with a controlled descent into the planet's atmosphere. By that time, its twelve instruments had streamed data for more than thirteen years, revealing hitherto unknown facts about this massive, gaseous planet and its rings; its huge moon Titan with its thick, murky atmosphere; the cold and geyser-rich Enceladus; and the entire Saturnian system.

Hubble Space Telescope

Like the concept of a massive space station, the idea of an enormous telescope orbiting the Earth, far above the interference of the atmosphere, took root well before the space age. As early as 1946, astrophysicist Lyman J. Spitzer of Princeton University proposed this radical new way of observing the universe. It gained further credence with Dr. Fred Whipple, the director of the Smithsonian Institution's Astrophysical Observatory at Harvard University. In hearings before Congress in 1959, Whipple—well known to the American people as a leading space exploration proponent—spoke about the benefits of a big telescope untethered to any fixed point. Six years later, the National Academy of Sciences Space Science Board advocated the launch of an observatory on the mighty Saturn V rocket. NASA followed its advice, putting the Apollo telescope mount aboard the 1973 Skylab mission. But to realize Spitzer's and Whipple's dream of a full-scale space observatory required decades of persuasion and preparation, vast outlays of money, and a mastery of technical complexity that rivaled the greatest engineering projects of the space age.

The name given to the first of these projects befitted its magnitude. The Hubble Space Telescope (HST) honored Edwin P. Hubble (1889–1953), the director of the Mount Wilson Observatory, located in the San Gabriel mountains north of Pasadena, California. Hubble's astronomical sightings transformed twentieth-century cosmology. He not only verified Albert Einstein's theory of an expanding universe, but also demonstrated that the cosmos held not one, but a vast profusion of galaxies, for which he developed a classification system.

Design work on the Hubble began during the 1970s, which slowed due to erratic funding, but then accelerated after NASA gave it a firm go-ahead in 1978. Its engineers planned a spacecraft of unprecedented proportions and capabilities: a silver canister the size of a railroad car, it measured 43 feet (13 meters) long and 14 feet (4.2 meters) in diameter; weighed 13.3 US tons (12.1 metric tons); and at its heart, possessed an immense, 94-inch (239-centimeter) primary mirror honed to precise tolerances, able to offer unmatched visual light observations of stars, galaxies, and extragalactic objects as far as 15 billion light years away.

Practical realities took some of the gloss off of this extraordinary machine. By early 1983, NASA Administrator James Beggs admitted to Congress that the project had fallen behind schedule and faced serious cost overruns. The House and Senate agreed to more funding, but on condition that Beggs cut expenditures where possible and tighten Hubble's management.

Hubble Space Telescope (HST)

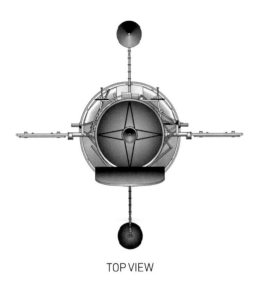

TOP VIEW

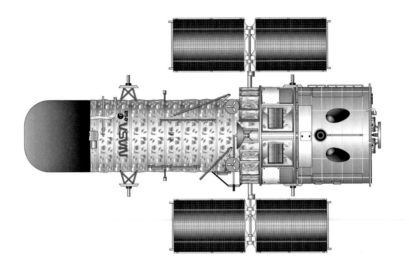

FRONT VIEW

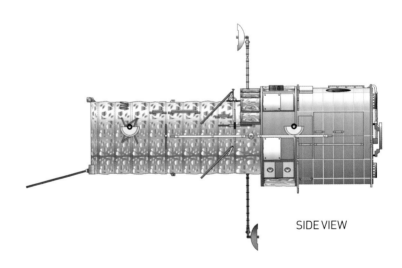

SIDE VIEW

meters

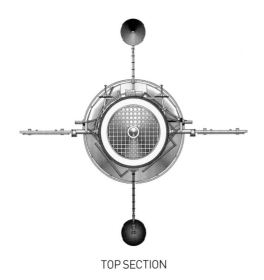

TOP SECTION

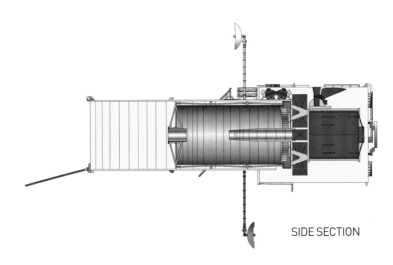

FRONT SECTION

SIDE SECTION

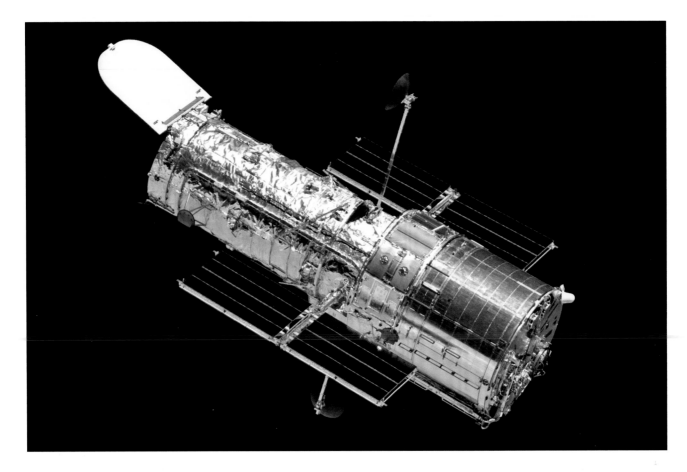

These demands actually catalyzed the project: NASA Goddard Space Flight Center assumed overall control of Hubble, in addition to providing oversight of the Space Science Telescope Institute at Johns Hopkins University, which directed the project's science operations. Lockheed Missiles and Space Company in Sunnyvale, California, the prime contractor, began to make better progress, and NASA signed an agreement with the European Space Agency (ESA) enabling their scientists to share access to the telescope. Unfortunately, spending on Hubble continued to inflate, in the end rising to over $2 billion.

While the technicians at Lockheed fabricated HST and conducted ground tests, the telescope lost its only means of transit into space. The *Challenger* accident in January 1986 grounded the shuttle fleet, relegating the nearly completed instrument to storage. There it remained until April 24, 1990, when *Discovery*, on mission STS-31R, lifted the hulking spacecraft into orbit—possibly the most significant nonhuman cargo carried by the shuttle to that point, or perhaps ever. Prior to the launch, astronauts and mission planners in Houston underwent countless rehearsals for the moment when they brought the HST to life. When this time came, Americans watched

anxiously as astronomer Steven Hawley grasped the Hubble with the shuttle's Canadarm and maneuvered this gigantic but delicate machine out of the cargo hold before releasing it.

The ease with which it all happened almost seemed anti-climactic—until the telescope's first observations less than a month later. At that point, NASA's leaders admitted the unthinkable: that Hubble failed to focus on distant objects, especially on stars. An investigation traced the problem to a Lockheed subcontractor who inadvertently fed incorrect data into the computer-controlled machine that ground Hubble's primary mirror, resulting in a malformation of its shape. Not only that, but the HST oscillated as it orbited—the result of its solar arrays heating and cooling as they passed in and out of the Earth's shadow—which confused its celestial pointing system.

NASA suffered withering criticism for the Hubble debacle (including mockery on the David Letterman talk show), but responded with plans for a rescue mission that offered no guarantee for success. Yet, engineers on the project found reason to hope; Hubble's mirror had been cut incorrectly, *but uniformly*, leaving open the possibility of restoring its sight by raising the surface of the mirror by as little as 2 micrometers.

Three years and eight months after Hubble's initial release, on December 2, 1993, *Endeavour* (in mission STS-61) sent a seven-person crew on a daring rescue. To make Hubble function as it should, the astronauts anticipated a dangerous mission. If they failed, no one could blame it on lack of preparation. They underwent exhaustive practices at Johnson Space Center with such components as the corrective optics, a suite of new computers, a replacement for the wide-field planetary camera, and new solar arrays. No previous extravehicular activities (EVAs) lasted as long or posed such high risks as the ones they embarked on.

After capturing the towering spacecraft with the Canadarm and locking it upright in the shuttle's payload bay, the crew grappled with it for five long days, from December 4 to 8, using more than two hundred tools that weighed an aggregate of 14,400 pounds (6,530 kilograms). The results won worldwide attention. Perhaps more influential than any astronomical instrument since Galileo's, Hubble's corrected vision attracted the public's interest in part due to the drama of its resurrection, and in part because of its celebrity on the internet. It also wasted no time in revising much of the prevailing orthodoxy about the universe.

Encouraged by the remarkable success of the first repair mission, three more followed in February 1994, December 1999, and March 2002. They extended Hubble's range beyond visible light to near infrared, replaced four of its six gyros, and installed updated solar panels and the advanced camera for surveys. The final trip to HST in May 2009 required five spacewalks and may have been the most ambitious visit of all: the astronauts mounted the cosmic origins spectrograph and wide field camera 3, repaired the space telescope imaging spectrograph and the advanced camera for surveys, and added a new science computer.

The Hubble Space Telescope continues in service twenty-eight years after its launch. During its long career, it penetrated far into a small portion of the universe called the Hubble Deep Field, viewing about three thousand galaxies—some up to 10 billion years old—for clues about the formative period of the cosmos. It observed the most distant stellar explosion seen yet, a supernova that detonated about 10 billion years ago (observations from which suggest that gravity decelerated the expansion of the universe after the Big Bang). Hubble identified the cataclysmic events that occur unseen at the heart of some galaxies, revealing them to be super massive black holes capable of swallowing light and making themselves invisible. The space telescope also linked planet formation to the pancake-shaped disks associated with some young stars, the frequency of which suggest that planetary births happen routinely.

Finally, Hubble enabled astronomers to calculate the age of the universe at between 12 and 14 billion years old.

The Hubble telescope will continue to transmit images until its instruments fail, probably after 2020. Meanwhile, in 2018 the James Webb Telescope will begin its career as Hubble's successor.

Compton Gamma Ray Observatory

In planning for its Great Observatories program during the 1980s, NASA assigned each telescope a mission based on the portion of the light spectrum that it covered: Hubble, visible light; Chandra, x-rays; and Spitzer, infrared. As its name suggests, the gamma ray observatory covered that specific band.

NASA named it the Compton gamma ray observatory shortly after the shuttle *Atlantis* launched it into orbit on April 5, 1991, on STS-37—about one year after Hubble, making it the second of the Great Observatories in space. It honored physicist Arthur H. Compton of the University of Chicago, who shared the Nobel Prize in physics with C. T. R. Wilson in 1927 for their explanation of changes in the wavelength of x-rays as they collided with electrons in metals (proving that electromagnetic radiation existed both as waves and particles). This discovery contributed substantially to the gamma ray observations made by Compton's instruments.

Much rode on Compton's success. NASA's reputation suffered with the blurred images being sent home from Hubble, and another embarrassment only a year later promised bad publicity and possibly funding consequences for the space agency. The Compton team held its breath as the shuttle astronauts unpacked this 17-ton (34,440-pound, 15,623-kilogram) behemoth—the shuttle's heaviest payload, with the exception of the later Chandra Inertial Upper Stage combination. They worried, too, about the telescope's solar arrays, which in ground tests proved problematic. Surprisingly, they opened without incident. The normally reliable high-gain dish antenna, however—indispensable because it sent the science data back to Earth—failed to unfold. Mission controllers tried everything they could to move it, even jostling it with the Canadarm, but nothing worked. So, astronauts Jay Apt and Jerry Ross suited up and went on an unscheduled spacewalk. Since the mechanism itself seemed undamaged, Ross got permission to apply elbow grease. Bracing himself with his right hand, he shoved twice on the boom that held the antenna; with a third and fourth push, it started to budge; two more tries and it became freed. Ross and Apt then went to the far end of the boom, opened the antenna fully by hand, and locked it down. With that, the crew released Compton.

Spared a disabling slip-up like Hubble's, the new telescope began its mission to scan for gamma ray activity,

Compton Gamma Ray Observatory (GRO)

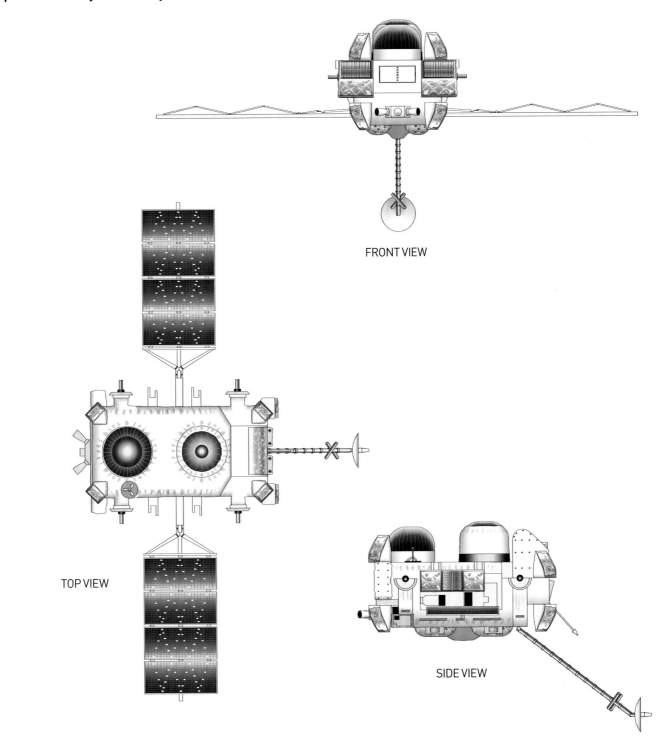

FRONT VIEW

TOP VIEW

SIDE VIEW

meters

Galileo (Jupiter Orbiter Probe)
With HGA partially deployed

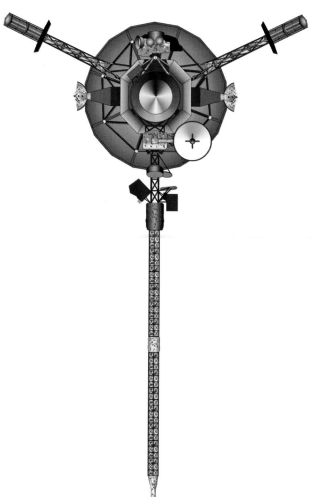

FRONT VIEW

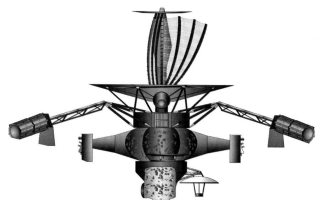

TOP VIEW

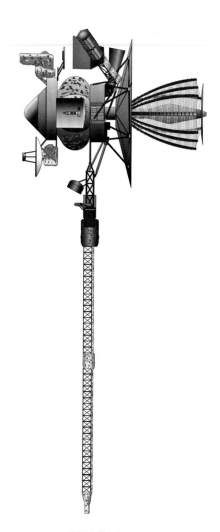

SIDE VIEW

0 1 2 3 4 5 meters

the highest-energy light known, but relatively rare; and rarer still for the most powerful bursts. But with its four instruments—the burst and transient source experiment (BATSE); the oriented scintillation espectrometer experiment (OSSE); the imaging compton telescope (COMPTEL); and the energetic gamma ray experiment telescope (EGRET)—Compton harnessed ten times the sensitivity of past gamma ray missions.

The resulting observations helped to rewrite some fundamental assumptions about the universe. BATSE, for instance, offered evidence that gamma ray emissions did not occur only in the Milky Way, as previously suspected, but all over the universe, even at its most distant points, commonly accompanying the death throes of massive stars or black holes. EGRET identified a new category of active galaxies dominated by supermassive black holes, responsible for the bulk of gamma ray flareups, with the power of over 100 million electron volts. And because of Compton's data, scientists believe that gamma ray discharges represented the strongest explosions in the heavens. Compton remained active for more than nine years. Mission controllers conducted an orbit reboost maneuver in 1993 that extended its life for five years beyond its initial voyage. But when a control gyroscope failed in late 1999, managers of the project decided to end it by atmospheric reentry, which occurred over the Pacific Ocean in June 2000. Meanwhile, during its service life, Compton recorded a total of over 2,600 gamma ray bursts.

Galileo

In the pursuit of planetary knowledge, one initial success often leads to a sequence of explorations, each elaborating on the one that went before. In this vein, the flyby missions of Pioneers 10 and 11 acted as catalysts for scientists to learn more about the mysterious planets lying beyond Mars and the asteroid belt. Pioneer 10 made the first contact, flying closest to mighty Jupiter in December 1973, after which it journeyed to Saturn. Next came Pioneer 11, which made its nearest approach to Jupiter a year later, and then also traveled to Saturn. The succeeding encounters with Jupiter occurred when Voyagers 1 and 2 collectively made observations of the solar system's four gas giants, starting with Jupiter in 1979 and then passing Saturn, Uranus, and Neptune. Yet, these missions only tantalized researchers. Well before these spacecraft reached their destinations, NASA planners envisioned a probe that not only orbited one of these vast worlds, but actually penetrated its interior. The space agency called it the Jupiter orbiter-probe and Congress approved it in 1977. But getting the go-ahead proved to be far easier than actually achieving this leap in capability. As conceived, the spacecraft consisted of two parts: an orbiter

designed and fabricated by the Jet Propulsion Laboratory (JPL) and a probe contracted to Hughes aircraft. By 1981, NASA already sank $300 million into the project.

To cut the time of Jupiter orbiter's long commute, JPL relied on a direct route to its target, accomplished by a shuttle launch, followed on orbit by a second firing from a new, three-part inertial upper stage (or IUS, developed originally by the air force as a two-stage booster for vehicles flying on its Titan missile). As NASA waited for the more powerful IUS, program costs kept rising, reaching about $650 million by the mid-1980s. Worse still, the air force abandoned the IUS then under development, leaving NASA no choice but to turn to the higher-energy Centaur upper stage. But this avenue also closed after the *Challenger* accident in 1986, when, in an effort to protect future astronauts, the space agency banned the volatile Centaur from being launched from the shuttle.

This decision left the project with the standard, two-stage IUS, which lacked enough thrust to send the spacecraft straight to Jupiter. Fortunately, JPL developed a workaround. To gather energy for the trip, project engineers devised a gravity-assist scheme involving Venus and Earth flybys, following which the vehicle headed for Jupiter. Unfortunately, rather than the original three-and-a-half-year flight plan, the JPL method required a six-year journey. One other factor threatened to inhibit the project: a lawsuit (eventually dismissed) challenged the environmental safety of the spacecraft's 22 kilograms of plutonium dioxide in its two radioisotope thermoelectric generators (RTGs)—the spacecraft's electrical power source.

At last, thirteen years after proposing it, NASA announced October or November 1989 as the launch window for the Jupiter project—by which time program costs increased to about $1.3 billion. Even so, now renamed for the celebrated Italian mathematician and astronomer Galileo Galilei (1564–1642), it went into orbit aboard the space shuttle *Atlantis* on October 18, 1989. Propelled by the two-stage IUS, it flew to Venus, passed Earth twice, and then in December 1992 turned its sights on Jupiter—meanwhile, observing two asteroids (Gaspra and Ida) as well as the comet Shoemaker-Levy.

Galileo arrived at the planet on December 7, 1995. Resembling a bowl with handles on each side, it measured 20.2 feet (6.15 meters) long from the top of the low-gain antenna to the bottom of the probe and weighed (at launch) about 5,648 pounds (2,562 kilograms). Galileo distinguished itself as the first space vehicle that stabilized both by a slow rotation of part of its body and a nonspinning part guided by gyroscopes. Among its eleven instruments, the orbiter's rotating part held equipment that studied charged particles, magnetic fields, and cosmic and

Jovian dust; the nonmoving segment contained instruments that required a fixed position, such as cameras and spectrometers.

The Galileo probe measured only 34 inches (86 centimeters) tall and weighed about 750 pounds (339 kilograms). Once at Jupiter, controllers released it, and it flew unpowered for five months (during which time the orbiter crossed paths with the moons Io and Europa and made a very low approach to the planet itself). When the day of actual contact arrived on December 7, 1995, the probe—flying at 106,000 miles per hour (170,000 kilometers per hour)—initiated aerodynamic braking, opened its 8-foot (2.5-meter) parachute, and for almost an hour transmitted data as it fell deeper into the planet's atmosphere. It was protected by aeroshells that blocked the heat generated during the descent, but high temperatures finally destroyed its electronics, rendering it lifeless. But before signing off, it sent home data about sunlight, temperature, pressure, winds, lightning, and the composition of the atmosphere.

About an hour later, the orbiter fired its main engine and fell into orbit around Jupiter. Its first revolution took seven months, followed by thirty-four more circumnavigations, the last beginning in November 2002. In its travels, it passed Europa eleven times, Callisto eight, Io seven, Ganymede six, and Amalthea once. The primary mission lasted through December 1997, after which Galileo pursued its first mission extension: a two-year voyage devoted to closer observations of Europa, Io, and Jupiter. Then the Millennium Mission from 2000 to 2003 featured flybys of Jupiter's major moons, measurements of the planet's magnetosphere (conducted in concert with the Cassini probe on its way to Saturn), a flyby of the moon Amalthea, and finally, on September 21, 2003, an intentional plunge toward the Jovian center and burn-up as it encountered the planet's dense atmosphere.

During its eight years in the grip of Jupiter, Galileo made several prominent discoveries. It found thunderstorms with lightning strikes one thousand times more powerful than those on Earth. It found evidence suggesting that liquid oceans flow beneath Europa's icy surface, and on Ganymede and Callisto as well. Researchers also found that Ganymede, Europa, and Io all possess metallic cores. Alone among all other moons in the solar system, Ganymede, like the Earth, generates a magnetic field. Io exhibited volcanic activity up to one hundred times more energetic than the Earth's. Scientists surmised that Jupiter's ring system evolved from dust dispersed when interplanetary meteoroids crashed into its four small inner moons.

Just as Galileo followed Pioneers 10 and 11 and Voyagers 1 and 2 in their pursuit of Jupiter, in 2016 the Juno spacecraft arrived at the planet, elaborating on the explorations that preceded it.

Giotto

The year 1986 marked a crossroads in the international pursuit of spaceflight. In a sense, it represented the start of a rebalancing of space leadership, from US predominance in the period during and after Apollo to a tripartite model shared by NASA, the Russian Federal Space Agency, and the European Space Agency (ESA). This relationship would become increasingly four-sided with the rise of the Chinese space program.

The realignment began in January 1986, when the Americans experienced the heaviest loss in space to date: the orbiter *Challenger* and her seven-person crew died shortly after liftoff in a violent explosion, leaving the US unable to launch astronauts into space during the shuttle's thirty-two-month shutdown. But as the US program suffered its worst setback, in March of that year the Soviets sent Mir—the world's biggest space station at the time—on a historic fifteen-year journey that gained invaluable knowledge about long-duration spaceflight. And during the same month as Mir's debut, the Europeans stepped onto the international stage by achieving their first deep-space mission with the Giotto probe, aimed at Halley's Comet. First recorded as early as 239 BC as a brilliant illumination across the night sky, and returning every seventy-five years since, Halley's represents perhaps the most striking astronomical occurrence visible with the unaided eye. During this voyage, Giotto achieved the first close encounter with any comet—indeed, not just any comet, but the most famous one of all.

Befitting a project that sought to unveil a mystery that has beguiled onlookers since antiquity, Giotto took its name from the early Italian Renaissance painter Giotto di Bondone, whose witness of Halley's Comet in 1301 so impressed him that he recast it as the star of Bethlehem in his painting, *Adoration of the Magi*.

For all its accomplishments, Giotto made an unassuming appearance: a squat cylinder with a broad belt of solar panels at its midsection. Below this portion, a round tray held its science equipment, and at the very bottom of the spacecraft technicians installed a bumper shield to protect Giotto from the debris of Halley's dust cloud. At the top, ESA engineers installed a high-gain antenna dish. Only a bit more than 5 feet (1.6 meters) tall and almost 6 feet (1.8 meters) in diameter, it weighed 2,116 pounds (960 kilograms) at launch and just 1,190 pounds—540 kilograms—when it reached its destination.

After its launch in July 1985 aboard an Ariane 1 rocket, Giotto approached its target about eight months later after a trek of more than 93 million miles (150 million kilometers). During the

Giotto (Halley Comet Probe)

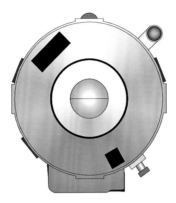

FRONT VIEW

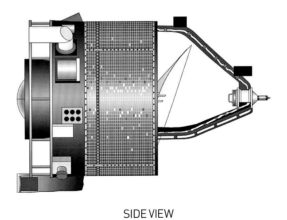

SIDE VIEW

0   1
meters

two hours leading up to its closest encounter with Halley's, it survived about twelve thousand dust impacts. Giotto attained its nearest point on March 13, 1986, at about 370 miles (596 kilometers) from the nucleus, but just before that moment a particle from the comet weighing about .035 of an ounce (1 gram) struck the spacecraft, sending it into a spin and temporarily ending its broadcast. Despite this blow, the ESA team quickly regained contact.

Images of the comet captured by Giotto's camera revealed an irregular body about 9.3 miles (15 kilometers) long and 4.3 to 6.2 miles (7 to 10 kilometers) wide. Other measurements found that water constituted 80 percent of the molecules that sprayed out of Halley's (at the rate of roughly 11 US tons per second). Dust measurements showed that 3.3 tons per second spewed out of seven distinct jets, consisting of large quantities of hydrogen, carbon, nitrogen, and oxygen.

After a long period of hibernation, Giotto received a wakeup call from ESA for one more mission: on July 10, 1992, it completed a 155.3 million-mile (215 million-kilometer) trip to the Grigg-Skjellerup Comet, where it made the world's closest cometary flyby to date at just 62 to124 miles (100 to 200 kilometers) from the nucleus.

On the next visit of Halley's Comet to Earth in 2061, people all over the world will again look at it with wonder, as they always have; but they will also comprehend it much better thanks to the observations of Giotto.

Ulysses

Like the mythical Greek king of the same name who, after the fall of Troy, wandered uncharted portions of the Mediterranean Sea for a decade, the Ulysses spacecraft also set forth on an epic adventure to unknown territory. By the time its travels ended in June 2009, it had journeyed for nearly twice as many years as the legendary Greek hero.

The modern Ulysses came about as a collaboration between the European Space Agency (ESA) and NASA to study the sun and its poles, and by inference other stars as well. Researchers wanted to discover not only more about the immediate solar environs, but also about the sun's influence on the heliosphere. Ulysses's mission planners hoped to illuminate such mysteries as the origins of solar winds and cosmic rays, the motions of high-energy particles, the sources of gamma ray bursts, the nature of interplanetary and interstellar dust, and the attributes of the sun's magnetic field. Since the sun operates on an eleven-year cycle within which its behavior swings from relative quiescence to high activity, Ulysses also assessed the impact of these solar seasons on the surrounding space environment.

ESA and NASA divided Ulysses's mission according to the instrument suites that each provided. The Europeans concentrated on equipment that detected the magnetic field, energetic particles, cosmic dust, and gravitational waves; the Americans on solar wind, low energy ions and electrons, cosmic rays, solar particles, and gamma ray bursts. Half of the satellite's ten instruments originated with the ESA countries and half with NASA.

The US space agency launched Ulysses aboard the shuttle *Discovery* (STS-41) on October 6, 1990, after a four-year delay due to the *Challenger* accident. The Jet Propulsion Laboratory (JPL) provided mission control; ESA designed and fabricated the spacecraft itself. Essentially a box with a high-gain antenna mounted on one face, it measured 10.5-by-10.8-by-6.9 feet (3.2-by-3.3-by-2.1 meters) and weighed 820 pounds (370 kilograms).

Ulysses started its voyage pointed toward Jupiter. Boosted by an inertial upper stage (IUS) and a payload assist module, it flew outward at the highest velocity of any spacecraft until that time. It reached Jupiter in February 1992 and, after a flyby, used gravity-assist to swing into a highly elliptical polar orbit of the sun. Ulysses first flew over the solar poles in 1994 and 1995, each pass typically taking between three and four months. Its primary mission ended in September 1995, but ESA and NASA authorities decided to lengthen its service. It observed the tail of the comet Hayakutake in 1996 and passed over the sun's poles again in 2000, 2001, 2007, and 2008.

Ulysses observed many tantalizing phenomena. After measuring the abundance of helium isotopes in interstellar gas (never before achieved), many researchers came to the conclusion that the universe lacked sufficient matter to destroy itself in a final collapse. And this mission also revealed that the magnetic field in the heliosphere—which lacks a cohesive pattern—behaves in a more complex fashion than thought previously; but, paradoxically, that the sun operates like a simple bar magnet in accomplishing its reversal of polar magnetism every eleven years.

Not long after Ulysses flew over the poles for the third time, its managers prepared to end the project. Because the spacecraft traveled in an elliptical orbit that carried it far from the sun, it relied not on solar energy, but on a nuclear-powered radioisotope thermoelectric generator (RTG) for electricity. Almost twenty years into the mission, the RTG ran down, shutting off the on-board heaters, resulting in frozen fuel lines. Mission control closed down Ulysses in June 2009.

Ulysses Probe

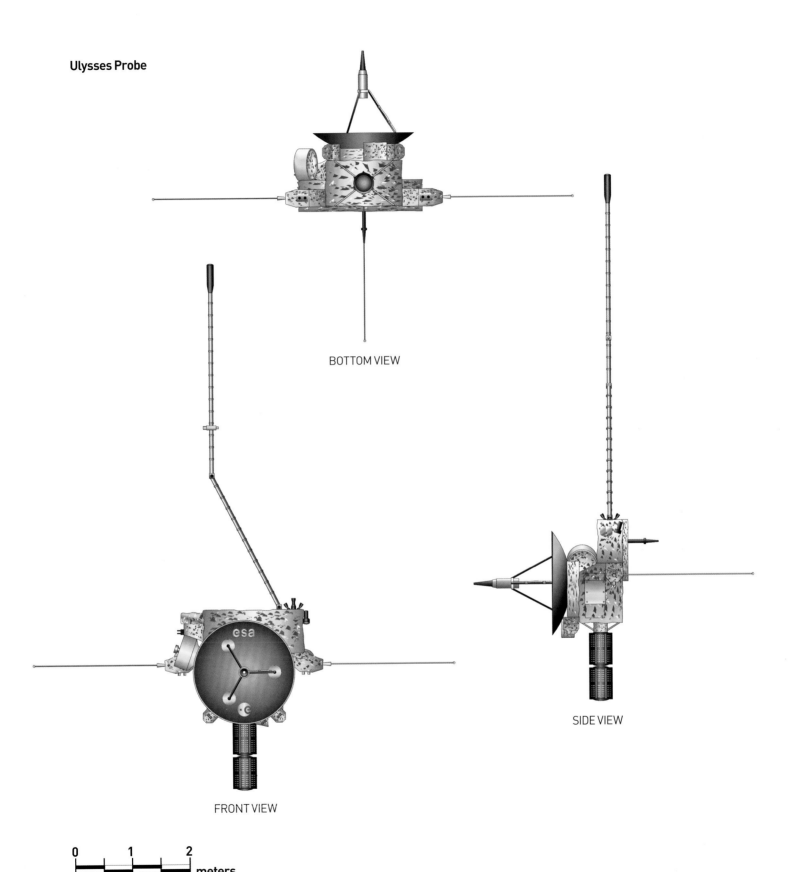

BOTTOM VIEW

SIDE VIEW

FRONT VIEW

0   1   2
meters

3

SPACE
EXPLORATION
AT A
CROSSROADS:
1997–2017

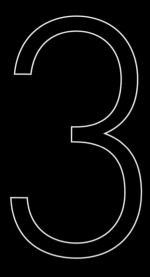

CAPSULES
Orion

After the shuttle *Columbia* broke apart during reentry on February 1, 2003, the Bush administration heeded one of the main suggestions of the subsequent *Columbia* Accident Investigation Board: to retire the shuttle after it completed the assembly of the International Space Station. In January 2004, President George W. Bush announced 2010 as its final year of service. In its place, he proposed a complex and ambitious program called the Vision for Space Exploration, which in one sense at least surpassed the shuttle, a spacecraft confined to Earth orbit. Bush called for a crew capsule and launch vehicle capable of sending astronauts into deep space. The president offered a timetable for the shuttle's successor: first flight in 2014 and a mission to the moon no later than 2020. Beyond that, he spoke expansively of the moon as a platform for eventual human visitations to Mars.

The optimism of Bush's speech soon met the grim budget realities brought on by the war on terror in Afghanistan and Iraq. NASA found itself paring down the president's grand scheme, supplanting costly new technologies with a system based on time-honored hardware. Constellation, as it came to be called, required two separate launchers: the Ares I, a rocket derived from the shuttle's solid rocket boosters; and the Ares V, a massive vehicle powered by five RS-68 engines (like those on the Delta IV missile), fueled from a huge external tank and supplemented by two additional solid rocket boosters. Ares I and V both used the Saturn V's J-2 second stage.

The creators of Constellation selected a spacecraft, the Orion command and service module, to transport astronauts beyond Earth orbit. Like Ares I and V, Orion drew heavily from the past. It strongly resembled the cone-on-cylinder profile of the Apollo command and service module, only more spacious. It could carry six astronauts to the International Space Station (ISS) and four to the moon. During lunar voyages, it reenacted Apollo's lunar orbit rendezvous architecture: as Orion orbited the moon, a lander

NASA Orion MPCS (Multi-Purpose Crew Vehicle)
ESM (ESA Service Module)

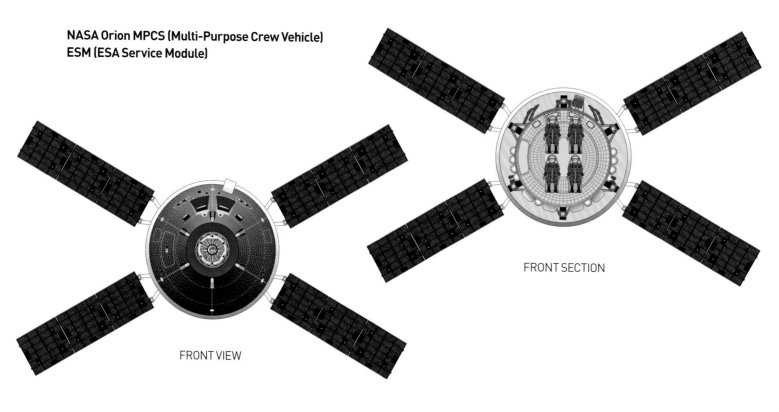

FRONT SECTION

FRONT VIEW

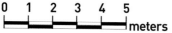

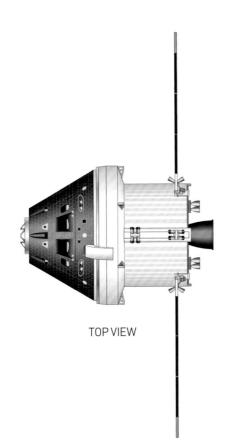

TOP VIEW

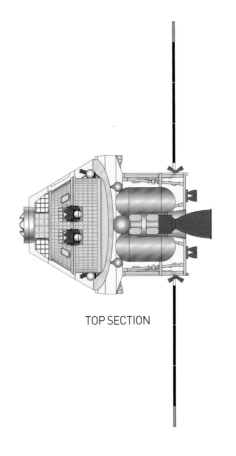

TOP SECTION

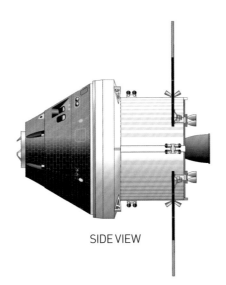

SIDE VIEW

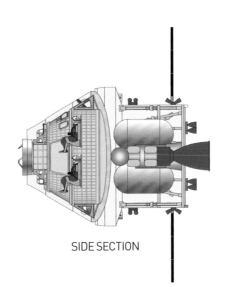

SIDE SECTION

Space X Dragon 2

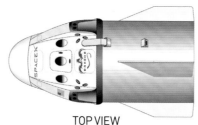

FRONT VIEW

TOP VIEW

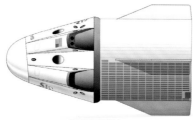

SIDE VIEW

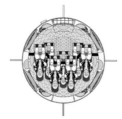

FRONT SECTION

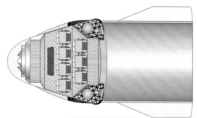

TOP SECTION

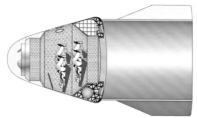

SIDE SECTION

0 1 2 3 4 5 meters

separated from it, flew to and touched down on the surface, remained there for up to seven days, and (using its upper stage) reunited with Orion for the trip home. The return to Earth both imitated and departed from Apollo. After jettisoning the service module, Orion's command module fell through the atmosphere on parachutes. But rather than a splashdown at sea, retrorockets and airbag cushions enabled the spacecraft to descend to points in the western United States for a ground landing.

During the first seven months of 2006, three aerospace firms—Lockheed Martin on one side and Northrop-Grumman plus Boeing on the other—bid for the Orion capsule project. NASA awarded the prime contract to Lockheed Martin on August 31, 2006, for $6.1 billion. But with the election of President Barack Obama in 2008, the means—less so the objectives—of the Bush space initiative changed. In a speech at the Kennedy Space Center in April 2010, President Obama outlined his plans, which included manned missions to asteroids, the moon, and even Mars. Obama also cancelled the Constellation project but spared the Orion spacecraft. In place of the Ares rockets, he proposed a more technologically advanced launch vehicle, which eventually became known as the space launch system (SLS). The government withdrew the contract with Lockheed Martin, but signed a new one that lasted until 2020 and obligated the company to build Orions for three missions.

Orion's command and service modules differed from Apollo most of all in volume. At 26.5 feet (8.1 meters) long and 16.5 feet (5 meters) in diameter, it measured about 10 feet shorter and 4 feet wider than Apollo. But more significantly, it offered a total of 691 cubic feet (19.5 cubic meters) of pressurized space compared to 366 cubic feet (10.4 cubic meters) for Apollo—almost twice the room. Orion weighed 56,985 pounds (25,848 kilograms)—surprisingly, about 10,000 pounds (4,500 kilograms) less than Apollo. Also, unlike the Orion associated with the Constellation project, the follow-on version landed at sea, rather than on land. And instead of the Constellation configuration that required three different Orion versions to achieve its varied tasks, NASA directed Lockheed to build the new Orion in one multipurpose design to accommodate all contingencies.

Testing of Orion began with a series of abort systems and splashdown recovery trials that occurred between 2008 and 2014. Then, on December 5, 2014, came the first orbital launch: a Delta IV heavy-lift rocket sent Orion twice around the Earth in about 4.5 hours. It reached an altitude of 3,600 miles (5,800 kilometers), after which it landed in the Pacific Ocean. Engineers assessed the spacecraft's parachutes, jettisoning equipment, on-board computers, and heat shield.

Following this flight, the next mission (scheduled for 2018) anticipates an unmanned seven-day voyage to the moon and back, followed by plans to launch the first crew on Orion, and by 2025, the space agency expects to send astronauts on an expedition to a nearby asteroid.

Resuming human-centered deep space missions—not attempted by NASA since Apollo 17 in 1972—did not come cheaply. Estimates of Orion costs alone (to 2023) total $20.4 billion, and this figure does not include the outlays for the enormous space launch system rocket still under development.

Dragon

When President George W. Bush and his administration decided in 2004 to cancel the space shuttle program after the *Columbia* accident, the future of US space exploration was in doubt. Predictably, the scheduled shuttle retirement in 2010 left NASA without the ability to send astronauts into orbit.

Industries responded quickly to the opportunity to vie for NASA's orbital business—focused on, but not limited to, resupplying the International Space Station (ISS) with materiel and eventually with crews. Among them, the Space Exploration Technologies Corporation (SpaceX) of Hawthorne, California (originated by entrepreneur Elon Musk), jumped out to an early lead. SpaceX had begun work on a space capsule called Dragon in 2004, and its Falcon 1 rocket made its first flight in 2006.

Just eight months after President Obama invited firms to compete for contracts, Dragon—launched from SpaceX's newer and more powerful Falcon 9—made its debut. In October 2012, Dragon became the first private spacecraft to dock at the ISS, where it brought supplies and returned cargo to Earth by splashdown. NASA signed agreements with SpaceX for fifteen ISS cargo missions through 2017, with at least six more from 2019 to 2024. The space agency agreed to pay SpaceX approximately $2.8 billion for these services. Although a major client, NASA represented one of many SpaceX customers; the company made forty-two total launches between June 2006 and July 2017, ten of which to the ISS for NASA.

In designing these new spacecraft, Dragon's engineers valued flexibility. In addition to the initial transport model, they developed a nearly identical crewed version (Dragon 2, carrying up to seven passengers), which SpaceX prepared for flight by using the cargo missions to the ISS to test human-rated components. A third and very similar Dragon variant called DragonLab will serve as a vehicle for technology demonstrations and science experiments.

Dragon's two sections consisted of a bell-shaped capsule with 388 cubic feet (11 cubic meters) of pressurized cargo space inside the cabin. Just below it, at its base, it held thrusters; a guidance, navigation, and control bay; and a heat shield. The second part, located underneath the spacecraft, had a cylindrical trunk containing the solar arrays and providing 494 cubic feet (14 cubic

**Space X Dragon CRS
Flight Configuration**

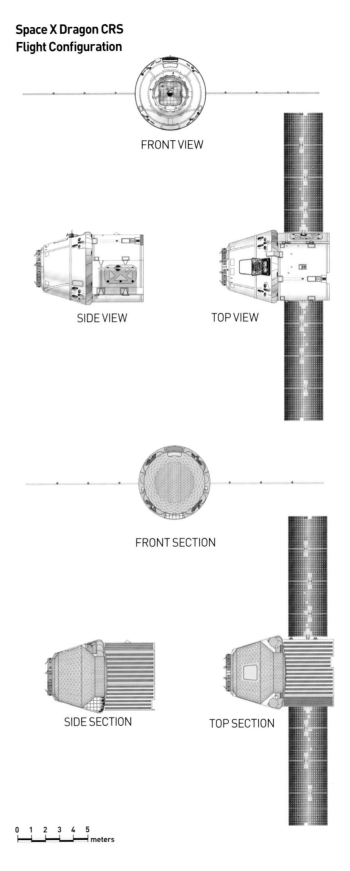

FRONT VIEW

SIDE VIEW

TOP VIEW

FRONT SECTION

SIDE SECTION

TOP SECTION

0 1 2 3 4 5 meters

Boeing Starliner (CST-100)

FRONT VIEW

TOP VIEW

0 1 2 3 4 5 meters

SIDE VIEW

FRONT SECTION

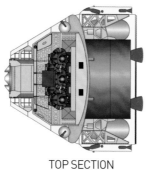

TOP SECTION

SIDE SECTION

0 1 2 3 4 5 meters

meters) of unpressurized payload capacity. In all, Dragon weighed 13,228 pounds (6,000 kilograms), measured 23.6 feet (7.2 meters) long, 12 feet (3.7 meters) in diameter, and offered 882 cubic feet (25 cubic meters) of interior space.

SpaceX won a go-ahead for Dragon 2 from NASA in 2014. For $2.6 billion, the company would develop the spacecraft, verify it in a test flight with an astronaut aboard, and then fly between two and six ISS crews to the station, with the scheduled first launch in 2018.

This agreement represented a new age in American spaceflight. If successful, it will end a seven-year period during which NASA relied exclusively on Russian launch vehicles to send its astronauts into space and start a period in which NASA relinquishes to private firms the role of launching US astronauts into orbit.

CST-100

SpaceX faced formidable competition in its response to President Obama's call in 2010 for commercial firms to bid on transport services for the International Space Station (ISS). Other relatively small firms such as Orbital Sciences and Jeff Bezos's Blue Origin also vied for these NASA contracts. But looming over all these fledgling companies stood the colossus of the aerospace industry with ambitious plans for its own spacecraft-launcher combination.

The Boeing Company started in this competition with big advantages. Not only did it (or the firms that it absorbed) embody a wealth of experience from the Apollo, shuttle, and ISS programs, it did not need to develop a new rocket. Because of a partnership that it formed in 2006 with Lockheed-Martin known as the United Launch Alliance, Boeing gained access to Lockheed's well-proven Atlas V. Flown since 2002, the Atlas V offered the heavy lifting power of a core booster with up to five strap-on solid rocket engines and a Centaur second stage.

Boeing responded in 2011 to President Obama's commercial launch invitation with a test model of an Apollo-shaped capsule that it called the Crew Space Transportation (CST)-100 Starliner. In October, the CST underwent wind-tunnel testing at NASA's Ames Research Center. The next year Boeing flew parachutes and air bags in drop tests, essential milestones because unlike Mercury, Gemini, and Apollo, this spacecraft landed on the ground, rather than in water. In 2012, NASA and Boeing agreed on the CST's basic layout, and in 2014 the company unveiled a full-scale mockup—including the complete interior layout—for the first time.

Boeing promised efficiency and modernity in its candidate. Because it avoided splashdowns, its engineers claimed that it offered reuse up to ten times. Its external shell consisted of a lightweight, nonwelded honeycomb structure, and its other

CAST Shenzou (Project 921-1)

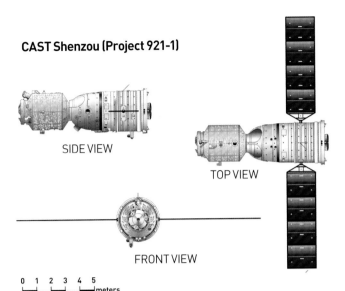

SIDE VIEW

TOP VIEW

FRONT VIEW

0 1 2 3 4 5 meters

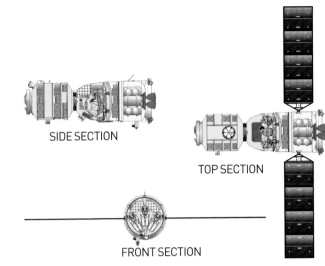

SIDE SECTION

TOP SECTION

FRONT SECTION

0 1 2 3 4 5 meters

features included autonomous docking, room for a crew of seven (NASA preferred the five-crew-plus-cargo configuration), automated controls backed up by manual piloting, and pleasing LED cabin lighting derived from Boeing's commercial aircraft. The CST measured 16.5 feet (5.03 meters) in length and 14.8 feet (4.5 meters) at its widest diameter.

Despite the milestones promised at the beginning of the project, the hard facts of development required technical and programmatic adjustments. The initial schedule predicted a pad abort test in 2013, an un-crewed orbital flight and a two-crew flight test in 2014, and the transportation of astronauts to the ISS in 2015. The pad abort test waited until October 2017, the un-crewed orbital flight test until December of that year, and the ISS mission (with one astronaut and one Boeing test pilot on board) until February 2018. The postponements occurred because Boeing needed to reduce the capsule's mass, resolve aerodynamics issues discovered in the launch and ascent tests of the CST-100 on the Atlas V, and finish software modifications resulting from new requirements requested by NASA.

Boeing proceeded with its work based on a $4.2 billion commitment from NASA in September 2014; SpaceX received about $2.8 billion at the same time. Whether these companies or other ones prevail, NASA authorities look forward to the results. In the end, choosing winners will free the agency from the predicament of astronaut transit, a situation bedeviling it since the shuttle's retirement in 2011.

Shenzhou

To judge by its string of successes in recent years, the Chinese human spaceflight program appears to be exceptionally coherent and focused. Even so, it responded to the same tides of politics

and budget priorities as all the major space powers. Accordingly, a manned program known as Shuguang began as early as 1968, promising a taikonaut (Chinese astronaut) in space by 1973. But a lack of funding and insufficient political interest doomed Shuguang. Low-level activities continued during the remainder of the 1970s, until 1980, when all work ended due to cost cutting.

The Chinese National Space Administration (CNSA) picked up the baton in the mid-1980s as China saw other countries making dramatic strides, most notably the US with its Freedom space station, strategic defense initiative, and space shuttle operations; and the USSR and Russia, with the Mir space station and the Buran spacecraft. After considering more advanced technologies such as reusable vehicles and spaceplanes, the Chinese authorities decided in 1992 to proceed with a capsule patterned consciously after the time-honored Soyuz spacecraft, paired with a variant of the Long March (Changzheng, CZ)-2F missile, upgraded for human passengers. The Soviets abetted the Chinese effort by offering fellowships to twenty Chinese engineers and, more directly, agreeing to transfer technology and astronaut training techniques to CNSA.

Once the new capsule underwent design, fabrication, and ground testing during the 1990s, it entered a second phase, making four unmanned flights beginning in November 1999 aboard the now human-rated Long March-2F. Although Chinese President Jiang Zemin proudly christened the spacecraft with the name Shenzhou, or Vessel of the Gods, only its service module flew as a fully functioning unit for the initial test; the still unfinished orbital module went into orbit as a dummy. Following that flight, designated Shenzhou 1, in January 2001 Shenzhou 2 went into space carrying a dog, a monkey, and a rabbit to test the life support system. Its recovery remains uncertain. Shenzhou 3 followed in March 2002 with a mockup passenger instrumented to assess the physical effects of spaceflight on human physiology. Shenzhou 4 in December 2002 represented a full rehearsal for the upcoming manned flight, with a crewmember in the capsule through part of the countdown. The mission went according to plan, clearing the way for the first taikonaut in space—fully thirty-five years after the initial discussions in China.

The spacecraft mounted at the top of the Long March-2F prior to the human launch debut looked uncannily similar to Soyuz, only bigger. Like the Russian vehicle, it consisted of three main sections: a forward orbital module (spherical in Soyuz; semi-spherical in Shenzhou); a bell-shaped reentry module in the center; and a cylindrical service module aft. Shenzhou weighed 17,284 pounds (7,840 kilograms) and measured more than 30 feet (9.25 meters) long, over 9 feet (2.8 meters) in diameter, with a span across its solar arrays of 55.7 feet (17 meters).

With Shenzhou 5, China joined the US and Russia as the only nations with the capability to send people into space. Taikonaut Yang Liwei lifted off on October 15, 2003. He orbited the Earth fourteen times in twenty-one hours before making a successful landing. Six more Shenzhou missions followed, in which the crew size, time on orbit, and complexity of tasks multiplied quickly—in

essence, compressing decades of space accomplishments into eleven years. During Shenzhou 6, launched on October 12, 2005, two taikonauts embarked on a five-day journey, during which they inhabited the orbital module of Shenzhou for the first time. Initiated on September 25, 2008, Shenzhou 7 carried three passengers on a three-day mission, during which two of them conducted spacewalks.

Then a second set of breakthroughs occurred. During the mission of Shenzhou 8, the just-launched Chinese space station Tiangong-1 received a visit from two test dummies during an automated rendezvous and docking on November 3, 2011. The first human habitation of Tiangong-1 occurred on June 18, 2012, when the crew of Shenzhou 9—two men and one woman—went aboard for eleven days. Shenzhou 10 also carried two men and one woman to the station where, beginning on June 13, 2013, they spent twelve days performing technology and science experiments, as well as research related to human physiology. Finally, on October 16, 2016, Shenzhou 11 carried two crewmembers to the new Tiangong-2 space station, setting a Chinese endurance record with a thirty-three-day mission.

Despite the long gestation period leading up to manned spaceflight, the Chinese space program gained increasing

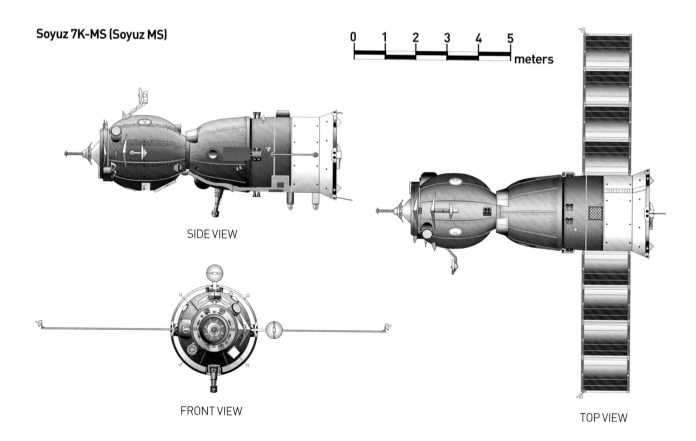

Soyuz 7K-MS (Soyuz MS)

0 1 2 3 4 5 meters

SIDE VIEW

FRONT VIEW

TOP VIEW

momentum once it began in earnest, and by the twenty-first century began to close the gap between itself and the Americans and Russians.

Soyuz MS (7K-MS)

The year 2016 represented an historic milestone in the history of spaceflight: it marked the fiftieth year since the Soyuz 7K-OK initiated the Soyuz series of capsules. Neither the 7K-OK, nor some of its successors proved to be wholly successful. But as a group, they demonstrated that a fundamentally sound design—the work of Sergei Korolev and his Special Design Bureau 1, later RKK Energia—offered engineers the opportunity to modify, modernize, and keep in service a family of spacecraft for two generations. From the outside, the Soyuz TM and TMA looked much the same, but the interior of the TMA reflected changes requested by NASA. As a client who often relied on the Soyuz for astronaut transport to and from the ISS, the US space agency asked for some practical improvements, such as adjustable couches to accommodate taller and heavier crewmembers. The Energia team also added improved parachute systems and the first glass cockpit—electronic flight instruments displayed on LCD

screens rather than on analog gauges—ever installed on an expendable space vehicle.

When the space shuttle retired in 2011, the entire responsibility for carrying astronauts and cosmonauts to the ISS and bringing them home fell for the foreseeable future to the Soyuz spacecraft. With that, Energia modernized the TMA and redesignated it as the Soyuz TMA-M. This entailed a general refurbishment of vintage equipment, in particular replacement of the old and heavy Argon computer with a new digital one, including digital avionics and displays. The TMA-M conducted twenty missions from 2010 to 2016.

Energia celebrated the fiftieth anniversary of Soyuz by unveiling a new model, the -MS ("modernized systems") in July 2016 (also known as the Soyuz 7K-MS). Although it looked like the TMA-M and other predecessors, the Soyuz MS heralded important modifications inside, related to electronics, crew safety, and flight control. Perhaps the most noteworthy involved the installation of a satellite navigation system, in preference to the old method that depended on six ground stations for orbital observation. Additionally, the reliable Kurs rendezvous system underwent its first major refinement in thirty years as the Kurs-NA, which introduced greater computer

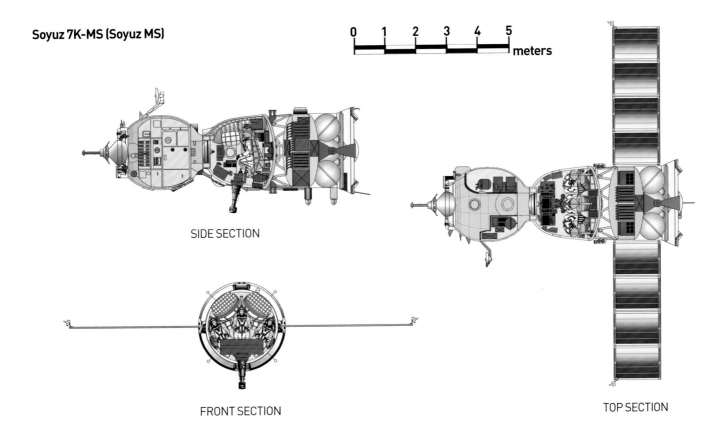

Soyuz 7K-MS (Soyuz MS)

0 1 2 3 4 5
meters

SIDE SECTION

FRONT SECTION

TOP SECTION

sophistication, along with smaller size, lighter weight, and lower power consumption. To compensate for the increased electrical needs associated with more advanced electronics, the MS's designers placed more cells on the solar arrays and added an extra battery to the four already on the spacecraft. In addition to these improvements, three satellites in a new communications system enabled the crew to connect to mission control in Moscow more than 80 percent of the time. Finally, responding again to NASA concerns, Energia reinforced the thin aluminum shell covering the habitation module on the MS to protect its three passengers from collisions with increasingly prevalent space junk, or with naturally occurring debris.

The first flight of the Soyuz MS took place on July 7, 2016, beginning with a two-day shakedown cruise. On the 9th, the crew docked with the ISS and remained there until October 30. Since then, the Russian space agency scheduled six more Soyuz MS trips to the ISS before the end of 2017, carrying fourteen passengers. They anticipate further missions up to Soyuz MS-12 in March 2019, with the option to extend them to the mid-2020s. By that time, Roscosmos, the Russian government's mirror to NASA, plans to unveil the next generation of manned Russian spacecraft.

SPACEPLANES
SpaceShipOne and Two

SpaceShipOne differs from most other advanced aerospace vehicles in that it originated not at a government research center, a university, or an aerospace corporation, but at the remote Mojave Airport, a quiet airstrip in California's high desert. Cars passing Mojave on Route 58 often see jetliners packed tightly on the desert sand, the byproduct of slumps in the airline business. Far removed from the main industrial infrastructure, Mojave drew inspiration from a nearby source. Driving east on Route 58, the Edwards Air Force Base exit comes into view in less than an hour. There, mounted on a pole outside of NASA's Armstrong Flight Research Center, stands a full-scale replica of the famed X-15 aircraft, flown in 199 test flights at Edwards from 1959 to 1968. What NASA and the US Air Force accomplished during the 1960s with the X-15 happened again in Mojave about forty years later. This time, however, a private firm rather than a federal agency designed, fabricated, and launched its own suborbital rocket plane with the ultimate intention of sending paying passengers on rides into space.

Burt L. Rutan (1943–), the engineer and entrepreneur who envisioned a privatized, updated X-15, began his career at Edwards,

where from 1965 to 1972 he worked as a flight test project engineer. He left to found a firm that made small planes for the home-built market, such as the Vari-Eze and the Long-EZ. But Rutan soon expanded his horizons. Notoriously impatient with the cautious pace of federally funded aerospace research, in 1982 he opened a new business called Scaled Composites, which he founded at the Mojave Airport. Rutan's firm specialized in business aircraft, research aircraft, and drones. Scaled Composites gained fame initially in 1986 when the Rutan-designed *Voyager* aircraft flew around the world nonstop without refueling.

The concept of a privately funded, suborbital spaceplane began to gel at Scaled Composites around 1994. Six years later, Rutan met Microsoft cofounder and space enthusiast Paul Allen over lunch, where Rutan sketched out on a napkin his concept of commercial access to spaceflight. In 2001, Allen pledged between $20–25 million to cover the costs of development. The resulting entity—Mojave Aerospace Ventures—combined Allen's pocketbook with Rutan's Scaled Composites.

Rutan and his group began the project with a first phase, Tier 1, to fabricate a spacecraft known as SpaceShipOne, a launch platform aircraft called White Knight, a hybrid rocket engine, and an avionics suite.

The flight architecture of this bold project corresponded closely to that of the X-15: a winged spacecraft slung under the body of a bigger mother ship, which took off together from an airport runway on suborbital missions. In the Rutan version, the pair climbed to 46,000 feet (15 kilometers), at which point the pilot of White Knight released SpaceShipOne, which glided briefly, then after rocket ignition climbed at 65 degrees. At the end of the burn, the spacecraft continued skyward at several times the speed of sound. Once it reached its desired altitude, SpaceShipOne descended, taking about twenty minutes to land. White Knight touched down a few minutes later.

White Knight—powered by two J-85 turbojets—looked at first glance like a jumble of slender trusses built from an Erector set. On closer inspection, its planform came into focus: long, thin wings formed into a "W" profile, dual tailplanes, a crew cabin, and a set of four wheels. Once airborne, it flew as high as 52,000 feet (16,000 meters) and carried a crew of two.

Scaled Composites fabricated SpaceShipOne's cigar-shaped fuselage from lightweight graphite-epoxy composite. They also incorporated short, wide wings with large vertical tailbooms, out of which jutted horizontal stabilizers. Reaction thrusters gave the pilot control in space. The main source of propulsion consisted of a hybrid rocket motor fueled by a mixture of solid hydroxyl-terminated polybutadiene and liquid nitrous oxide oxidzer. Inside, the sole pilot sat forward, with two passengers aft. White Knight

Scaled Composite SpaceShip One (TIER 1 Project)

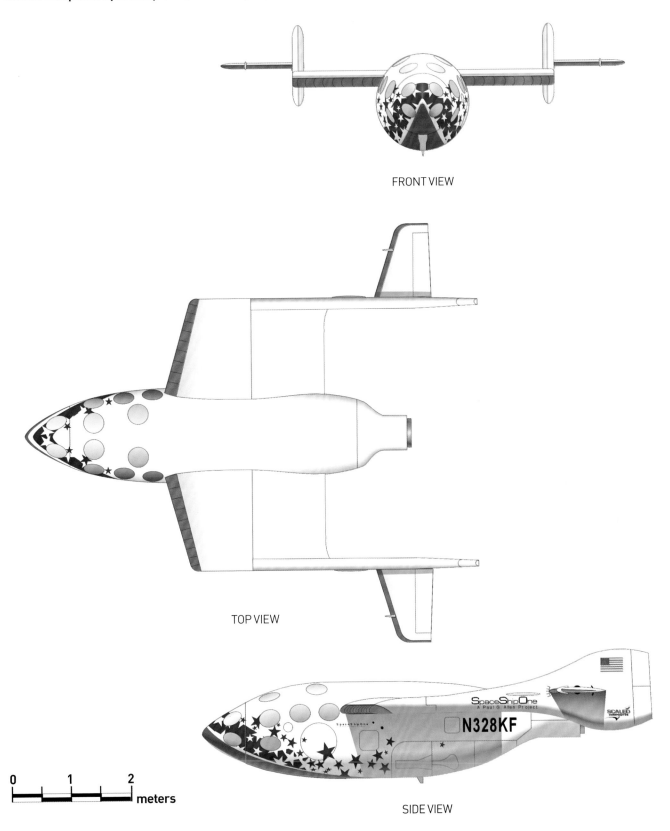

FRONT VIEW

TOP VIEW

SIDE VIEW

N328KF

0   1   2 meters

Scaled Composite Model 339 SpaceShip Two
Virgin Galactic VSS Unity

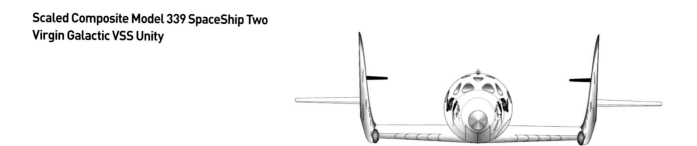

FRONT VIEW

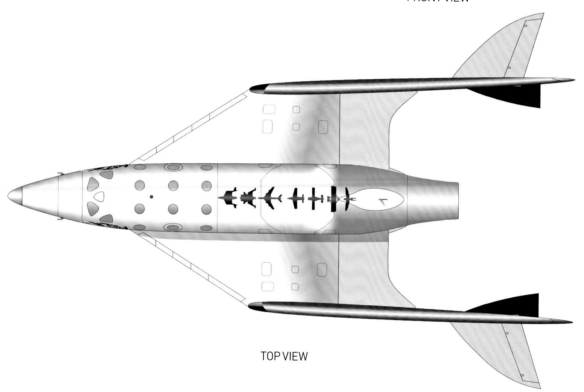

TOP VIEW

SIDE VIEW

meters

and SpaceShipOne shared the same forward fuselage design and avionics system.

SpaceShipOne weighed (loaded) about 7,920 pounds (roughly 3,600 kilograms), measured 28 feet (8.53 meters) long, 5 feet (1.52 meters) in diameter, with a wingspan of 16.5 feet (8.05 meters). It developed a maximum speed of 2,170 miles per hour (3,518 kilometers per hour) and climbed as high as 367,360 feet (112,000 meters). White Knight's long wingspan stretched 82 feet (25 meters) and it carried payloads of up to 8,000 pounds (3,629 kilograms). At takeoff, the aircraft weighed 18,960 pounds (8,600 kilograms).

During the flight research phase, White Knight underwent twenty-three solo trials from August 2002 to May 2003, before the first captive test with SpaceShipOne on May 20, 2003. After more than a year of flights preparatory to the ultimate goal, on June 21, 2004, South African pilot Michael Melville flew SpaceShipOne to an altitude of just over 328,084 feet (100,000 kilometers)— reaching the internationally recognized threshold for space. A little more than three months later, on September 29, 2004, Melville pushed the aircraft up to 338,000 feet (103,000 kilometers), ignoring advice from the ground to abort the mission when the spacecraft went into a roll as it climbed toward space. Finally, on October 4, 2004, pilot Brian Binnie flew SpaceShipOne to 367,454 feet (112,000 kilometers), breaking the X-15 altitude record of 354,200 feet (108,000 kilometers) set by NASA in August 1963.

More than that, in the last two flights the Allen-Rutan team satisfied the terms of the Ansari X-Prize, a global challenge made by Iranian-Americans Anousheh and Amir Ansari. The Ansaris offered to pay $10 million to the first private party that successfully completed two piloted, suborbital flights within two weeks of each other, carrying weight equivalent to two passengers. With much global media attention, Mojave Aerospace Ventures declared victory and collected the winnings.

But the saga did not end there. The news got the attention of Sir Richard Branson, founder of Virgin Atlantic Airlines and owner of a far-reaching business conglomerate known as the Virgin Group. Midway through 2005, Branson and Burt Rutan signed a contract merging the Virgin Group with Scaled Composites in a 70 percent / 30 percent split. They called the resulting entity the Spaceship Company. As stipulated then, in 2012 Branson's Virgin Galactic became the sole owner of the Spaceship Company when it acquired Scaled Composite's share.

Five years after this union, in 2010 the Spaceship Company began construction of a 68,000-square-foot plant in Mojave, where a growing work force began fabrication in September 2011 of five SpaceShipTwo spacecraft and three WhiteKnightTwo aircraft. Each

was approximately twice the size of SpaceShipOne and White Knight, but their designs did not differ significantly from their predecessors. SpaceShipTwo measured 60 feet (18 meters) long, with a wingspan of 27 feet (8.2 meters); its crew cabin, which accommodated two pilots, totaled 12 feet (3.7 meters) long, with a diameter of 7.5 feet (2.3 meters). The spaceplane seated six passengers.

For the future pleasure of experiencing a suborbital flight, Virgin Galactic collected roughly $50 million in deposits from over four hundred individuals; about 65,000 competed for the first hundred seats. Initially, they signed up for the adventure at a cost of $200,000 per head; it later increased to $250,000. SpaceShipTwo may also be contracted for scientific research by NASA and other entities.

Despite these brisk sales, Virgin Galactic's plans have become enmeshed in developmental setbacks. On July 26, 2007, three Spaceship Company employees died and three suffered injuries—all from shrapnel wounds—in a ground explosion that occurred during an oxidizer flow test. Then, on October 31, 2014, tragedy struck again when the first of the SpaceShipTwos disintegrated in flight due to a premature initiation of the descent system. It crashed, killing one pilot and injuring the other, who parachuted to safety. The rocket engine also caused concern. In May 2014, Virgin Galactic, dissatisfied with the contractor Sierra Nevada that built SpaceShipOne's powerplant, took over the research and testing of RocketMotorTwo itself. Virgin switched propellants, from Sierra Nevada's rubber-based original to a solid fuel, only to return to the initial formula in 2015.

Meanwhile, as the vehicles themselves continue to undergo flight testing in Mojave, the state of New Mexico paid for the construction of a modern terminal called Spaceport America, equipped with a 12,000-foot runway. Virgin Galactic signed a twenty-year lease as the site's main tenant. Located outside the town of Truth or Consequences, New Mexico, it awaits SpaceShipTwo launches and customers.

X-37B

Anticipating the eventual demise of the space shuttle as it approached its twentieth year of flight, NASA's leaders decided in August 1998 to solicit proposals for a demonstrator spaceplane to explore the engineering and science of cutting the cost of future space transportation. The space agency took this step realizing that unless new means of transportation arose, the aging and expensive space shuttle faced decades of supplying, servicing, and ferrying crews for the just-emerging International Space Station (ISS), whose construction also began in 1998.

Boeing won the competition for the demonstrator— designated the X-37—in December 1998, and the company and the space agency signed a four-year agreement in July 1999 with a cost-sharing provision committing the government for $125 million (including $16 million from the air force for its own technology experiments) and Boeing for $67 million. A second contract in November 2002 gave Boeing the go-ahead to fabricate two spacecraft: an approach and landing test vehicle (ALTV) and an unmanned X-37 orbital vehicle. This phase earned the manufacturer an additional $301 million.

But events in 2003 upended the planned X-37 in ways no one foresaw. As a result of the shuttle *Columbia* accident in February 2003, President George W. Bush's administration decided not only to terminate the shuttle in 2010, but to replace it with a new but technically conservative space capsule reminiscent of Apollo, and a heavy-lift launch system that resembled the Saturn V. With that, NASA no longer needed the X-37 and transferred it to the Defense Advanced Projects Agency (DARPA) in September 2004. DARPA took possession of the ALTV vehicle and conducted a series of captive carry and free flight tests until September 2006, after which the air force announced that it planned to continue the program under the designation X-37B orbital test vehicle (OTV).

In part, the air force structured the X-37 project as NASA had: as a testbed for the technologies needed for future spaceplanes, with a charter to explore such areas as advanced guidance, navigation and control, and high-temperature structures and more. But, in contrast, the USAF also conceived of the X-37B as a classified, reusable, experimental vehicle capable of sending payloads into orbit and returning them to Earth.

Boeing delivered two OTVs to the air force, each measuring 29.4 feet (8.9 meters) long, 9 feet 6 inches (2.9 meters) tall, with a wingspan of 14 feet 11 inches (4.5 meters). Intended originally to be launched from the space shuttle's cargo bay, the 11,000-pound (4,990 kilogram) X-37B actually went into orbit on board the Atlas V (version 501) rocket, boosted by a Centaur second stage. The X-37's body shape and delta wings drew inspiration from the aerodynamics of the shuttle orbiter, and they shared a similar lift-to-drag ratio.

Five OTV launches occurred between 2010 and 2017, although the exact purpose of each remains classified. Because of the information blackout, speculation has ranged widely, claiming that the X-37Bs pursue reconnaissance on other spacecraft, test space-based weapons, and conduct spy-sensor experiments. What the air force does reveal relates mainly to the liftoffs, landings, and the duration of each flight.

Whatever they may be accomplishing, the OTVs compiled lots of time in orbit. Numbers 1 to 4 entered space from Cape Canaveral Air Force Station, Florida, and all but OTV-4 touched down at Vandenberg Air Force Base, California. The first (OTV-1)

began its journey on April 22, 2010, traveled for 224 days, and made America's first autonomous orbital landing on a runway. The second (OTV-2) took off in March 2011 and came home after more than 488 days aloft. The third (OTV-3) followed in December 2012 (after a postponement due to an engine issue on the Atlas V), stayed in space nearly 675 days, and returned to Earth in October 2014.

The fourth OTV mission (OTV-4) went into orbit in May 2015 and remained there for almost 718 days, ending its mission at the Kennedy Space Center on May 7, 2017. For this flight, the air force identified two public objectives: to conduct experiments for NASA on roughly a hundred different materials exposed to the space environment and to conduct tests for the Aerojet-Rocketdyne Company of the Hall-effect thruster (an ion engine), intended for the advanced extremely high frequency (AEHF) communications satellite. Finally, OTV-5 broke the mold of the earlier launches, setting off in September 2017 aboard a commercial SpaceX Falcon 9 rocket.

In its design and operation, the X-37B represented little more than an automated space shuttle; but, depending on the data that the air force acquires from its flights, it might fulfill the original intention of the X-37B as a gateway vehicle to the full-scale spaceplanes of the future.

STATIONS
The International Space Station

The concept of a massive space station circling the Earth originated in modern times with an influential set of articles in *Collier's Magazine* that from 1952 to 1954 popularized the coming Space Age. It featured majestic images of orbiting spacecraft illustrated by artist Chesley Bonestell, with accompanying narratives by rocketry pioneer Wernher von Braun, and others. Among Bonestell's images, his austere yet beautiful drawings of enormous, wheel-shaped stations elicited excitement and a sense of wonder.

The space station ideal lingered and eventually became part of NASA's exploration program after its establishment in 1958. Nonetheless, a succession of political leaders and space agency administrators grasped that no one country—not the United States by itself—could muster adequate resources to pursue it independently. Its cost and complexity demanded a multinational effort. So, even before President Ronald Reagan announced in his State of the Union address in January 1984 his support for a heroically sized domicile in space, the White House directed NASA's leaders to contact six governments—the UK, Canada,

Boeing X-37B OTV (Orbital Test Vehicle) 2

FRONT VIEW

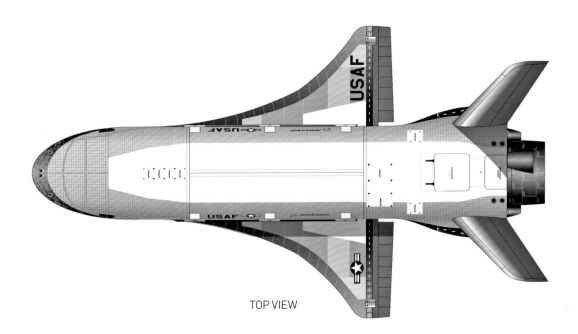

TOP VIEW

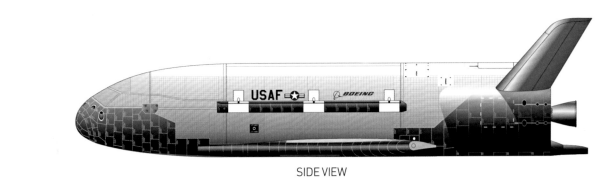

SIDE VIEW

0　1　2　meters

Germany, Japan, Italy, and France—to solicit their cooperation. For NASA Administrator James Beggs, who campaigned vigorously for Reagan's announcement, the station also offered an opportunity to jolt his agency out of the doldrums that had persisted since the glory days of Apollo.

The president underscored the importance of international collaboration when he subsequently sent Beggs on a tour of six cities—London, Ottawa, Bonn, Tokyo, Rome, and Paris—to meet with NASA's counterparts and ascertain their level of interest. Encouraged especially by his reception in Germany, Italy, and Japan, Beggs later informed his hosts that the president intended to submit a budget request for $8 billion for a US station, to be completed by the early 1990s. He also reiterated Reagan's desire to expand and improve the project through global partnerships.

But nothing in these early steps foretold the tortured series of events that stretched out the project's completion to twenty-seven years—nearly four times the gestation period hoped for by Reagan. And the cost wasn't $8 billion, as planned, but more than ten times that amount.

The big space station became problematic almost as soon as President Reagan announced it, however, and remained so into the twenty-first century. Part of the difficulty stemmed from politics and reflected a serious debate in Congress about whether the project really made wise use of limited national resources. Opposition gathered in 1985 almost as soon as James Beggs unveiled NASA's candidate design, called Space Station Freedom, a behemoth measuring 500 feet (152.4 meters) long, 360 feet (110 meters) tall, and carrying eight crewmembers. Even though the Japanese, European, and Canadian space agencies pledged their willingness to participate, many US legislators objected to the $8 billion proposal. NASA bowed to their pressure. By 1986, all the satellite fabrication facilities envisioned for Freedom disappeared from the plans, leaving only a laboratory. After the *Challenger* accident in 1986 and the subsequent hiatus in shuttle launches—which did nothing to strengthen the argument for space exploration in general—NASA further contracted the scale of Freedom to a crew of only four astronauts, confined to microgravity and life-science experiments.

The decline marched on during the early presidency of Bill Clinton, whose administration renamed the station Alpha and reduced its dimensions even more. But in 1993, NASA Administrator Daniel Goldin took a proactive step. He contacted

International Space Station (mid-2017 layout)

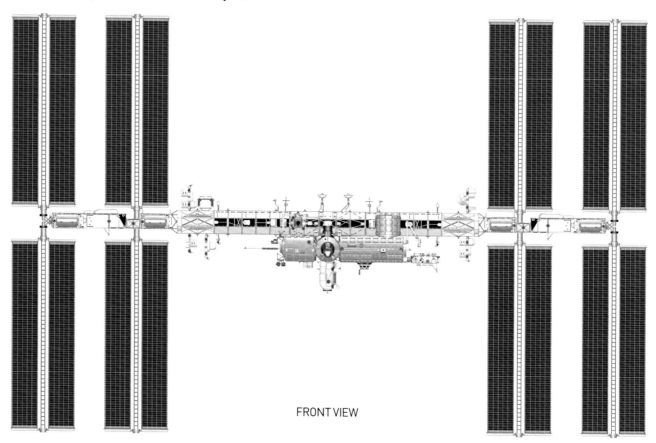

FRONT VIEW

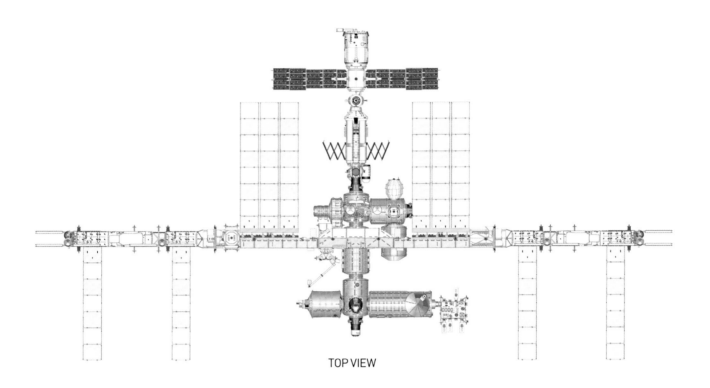

TOP VIEW

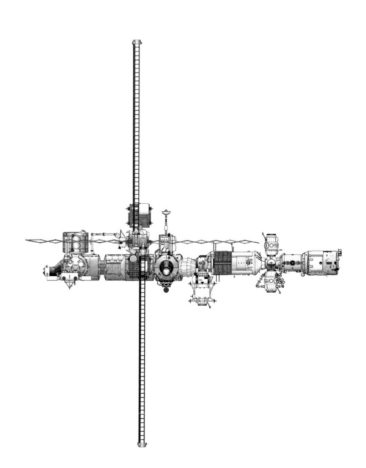

0 ___ 5 meters

SIDE VIEW

his counterparts in Roscosmos and persuaded them to combine their space station program with that of the United States. The Russian contribution ended the downward spiral of the US effort. Just as significantly, on the political front the Russian participation appealed to those US officeholders who approved of warmer relations between the two former adversaries.

Other American politicians, however, took exception to joining forces with a country that only recently represented (in their view) a menace to American institutions and security. And the ballooning space station budget (estimated by the early 1990s at about $17.4 billion) gave its opponents a second and more persuasive line of attack. A vote in the US House of Representatives in 1993 saved the project by a single vote. But just two years later, as memories of the Cold War receded, a partnership with the Russians seemed more plausible, and Congress approved a station budget capped at $2.1 billion annually.

But the controversy did not end there. By 1997 and 1998, it became clear that NASA breached the yearly spending limit and required an additional $7.3 billion over the baseline $17.4. Just as troubling, independent auditors determined that delays of between ten and thirty-six months in the delivery of key components rendered impossible the projected 2004 completion date. In response, America and its international collaborators focused initially on finishing in 2006; but the *Columbia* accident in February 2003 halted shuttle activity, and with it the delivery and assembly of new modules. Under these circumstances, the station partners chose 2010 as a more realistic target.

Officially, the International Space Station (ISS) came into being in January 1998 with the signing by fifteen governments—the United States, Russia, Japan, Canada, and the eleven member states of the European Space Agency—of the Space Station Intergovernmental Agreement. But the primary collaboration between the United States and Russia represented the indispensable ingredient in the evolution of the ISS. The US may have been the first (and the only) power to send people to the moon, but the Russians far outdistanced America in its space station experience. Roscosmos not only launched seven Salyut ("salute") stations from 1971 to 1986, it followed these in 1986 with Mir ("peace"), a dragon fly-shaped, 200,000-pound spacecraft, the biggest object in orbit to that time. Fortuitously, by the time Daniel Goldin approached Russia in 1993 with the hope of a partnership, Roscosmos already had made big strides toward Mir-2, but with little chance of funding due to the post–Cold War

chaos in the Russian economy. In effect, the ISS represented a merger of Mir-2 and Alpha. In pursuit of this amalgamation, Russian engineers and planners worked directly with their US colleagues, and Roscosmos pledged to add two new modules to the overall American architecture, increase the size of the crew quarters, lengthen the station's central corridor, and enhance power generation.

In addition to this technical symbiosis, in human terms both astronauts and cosmonauts benefited from early cross-fertilization of experiences: seven Americans flew aboard Mir between 1995 and 1998, allowing them to learn the techniques of prolonged spaceflight (up to six months); and seven Russians flew on seven shuttle missions during the same period, enabling them to become familiar with the space truck that built and nourished the ISS over many years.

The initial period of ISS on-orbit assembly occurred from December 1998 to January 2003, involving the integration of Unity, Zarya, Zvezda, Destiny, and the multipurpose logistics modules (MPLMs). The second growth spurt began with the return to flight by the shuttle *Discovery* after the loss of *Columbia*. During this phase, the ISS gained ESA's Harmony (Node 2) module in

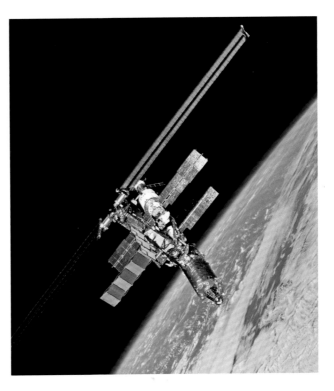

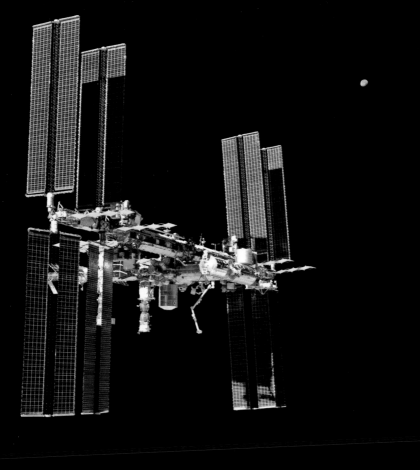

October 2007 and the Columbus laboratory in February 2008; in March and May 2008 and July 2009 the components of the Japanese Aerospace Agency's Kibo laboratory; in February 2010 the Italian Space Agency's Tranquility (Node 3) and seven-windowed observation Cupola; and finally, in February 2011, the modified Leonardo MPLM, returning to ISS as the permanent multipurpose module.

The main construction of the ISS lasted almost thirteen years, from 1998 to 2011, and represented perhaps as much a tribute to human endurance as to human ingenuity. Gradually, the massive structure took form. In all, forty-three assembly flights occurred, with a peak of seven visits in 2001. The completed ISS weighs about 882,000 pounds (400,068 kilograms), and at 356 feet (108.5 meters) long by 240 feet (73 meters) wide, it more than covers the dimensions of an American football field. By summer 2017, the ISS hosted 227 individuals and fifty-two different crews (or expeditions). The full cost of the ISS exceeded $100 billion, with the United States contributing $3 billion annually toward its continuing operations.

Originally scheduled to be closed and deorbited in 2010, the administration of President Barack Obama announced in 2014 an extension of its service life at least until 2024.

Unity

Although not exciting or especially innovative, America's initial contribution to the world's first multinational home in space served an indispensable purpose. Packed into the shuttle *Endeavour*'s payload bay, the Unity module (also known as Node 1) served as the central hub onto which most of the major pieces of the early station docked. Fashioned out of aluminum with six berthing locations, the barrel-shaped Unity arose on the floor of the Boeing Company's facility at NASA's Marshall Space Flight Center in Alabama. It weighed 25,600 pounds (11,612 kilograms) and measured 17.9 feet (5.47 meters) long and 15 feet (4.57 meters) in diameter. In addition to Unity, *Endeavour* also carried a crew of four astronauts and a cosmonaut trained to bring the new station to life. To do so, the shuttle needed to rendezvous with Unity's counterpart—a Russian spacecraft called Zarya ("dawn" in Russian) launched two weeks earlier.

As the shuttle and the Zarya spacecraft approached one another on December 6, 1998, Commander Robert Cabana took manual control of *Endeavour* and positioned it to within 10 feet of Zarya. At this point, astronaut Nancy Currie extended the shuttle's Canadarm, grasped Zarya, and drew it to Unity (anchored upright in the shuttle's cargo bay). She then used the robotic arm to line up the twenty-four pins and matching holes on the two spacecraft and, after some delicate maneuvering, locked the mechanism into place. Three spacewalks lasting a total of twenty-one hours followed during which the crew integrated Unity and Zarya's electrical and other vital systems. Following that task, on December 11 Cabana led his crew into Unity, where they installed lights and communications, and into Zarya, where they carried in tools and clothing.

Having joined together and partially outfitted the first components of the ISS, the crewmembers returned to *Endeavour* and on Sunday, December 13, left the newly awakened but empty International Space Station and returned home.

Zarya

True to the Russian Federal Space Agency's long tradition of modifying its spacecraft in increments and adapting older designs, its first contribution to the International Space Station (ISS) evolved from the Russian Transport Logistics Spacecraft (TKS), originally a ferry for the Almaz military space station. Called Zarya ("dawn" or "sunrise"), its name hinted at both the start of a new partnership between former rivals and the initiation of a massive new project in space. Known also as the functional cargo block (or FGB, following its Russian spelling), it served several purposes. First, its three ports rendered it a passageway between Russian and American components of the ISS, with a Russian docking point one at one end, an American at the other, and a berth for visiting Progress supply and Soyuz transport vehicles at a third. Functionally, Zarya acted as the ISS's initial infrastructure center. Its expansive solar arrays and six batteries generated 3 kilowatts of electricity; its sixteen external

Node 1 "Unity" + PMA 2

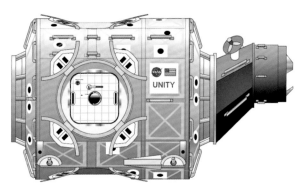

TOP VIEW

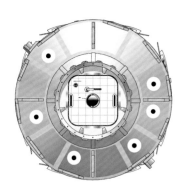

FRONT VIEW

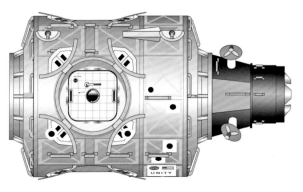

SIDE VIEW

meters

Zarya FGB (ISS Functional Cargo Block)

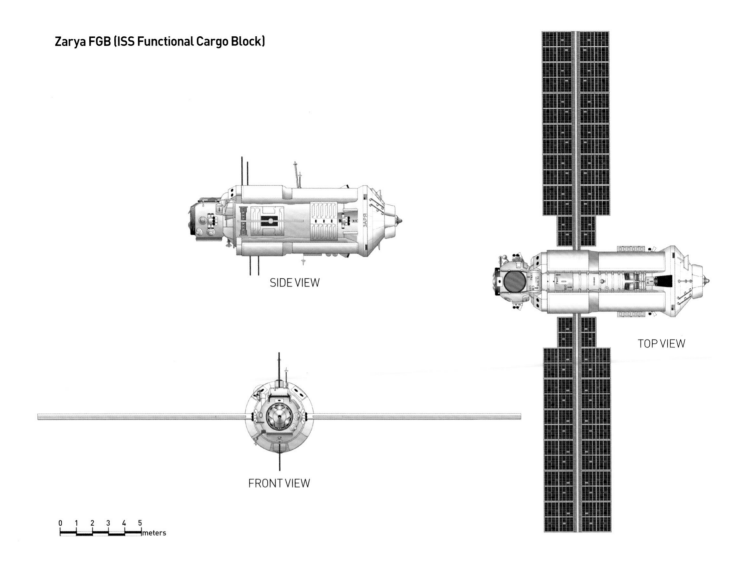

SIDE VIEW

TOP VIEW

FRONT VIEW

0 1 2 3 4 5
meters

tanks carried more than 13,200 pounds (6,000 kilograms) of propellant; it held two large engines, twelve small ones, and twenty-four steering jets (for orbital maneuvers); and it offered pressurized space that accommodated storage, as well as a crew compartment with windows for the astronauts and cosmonauts. A long, heavy, telescope-shaped object, Zarya weighed 42,600 pounds (19,323 kilograms) and measured 41.2 feet (12.5 meters) long and 13.5 feet (4.1 meters) in diameter.

Zarya took off from Baikonur Cosmodrome in Kazakhstan on November 20, 1998, lifted by a Proton three-stage, heavy launch rocket. It assumed a circular orbit at 240 miles where it awaited Unity, the first US contribution to ISS. Sixteen days later, the shuttle *Endeavour* arrived at Zarya with the American module in

its cargo bay. *Endeavour*'s manipulator arm captured Zarya, and after a delicate mating process, the pair released into orbit.

After an unexpectedly long twenty-month wait, on July 25, 2000, Zarya docked with the second Russian ISS component, Zvezda ("star"), the addition of which rendered the ISS suitable for human habitation.

Zvezda

With the mating of the first modules Zarya and Unity, the International Space Station (ISS) started to take shape in December 1998. Space station planners hoped for a quick expansion with the arrival of another Russian segment known as the Zvezda ("star") service module. But the months stretched

Zvezda ISS Service Module (DOS-7K #8)

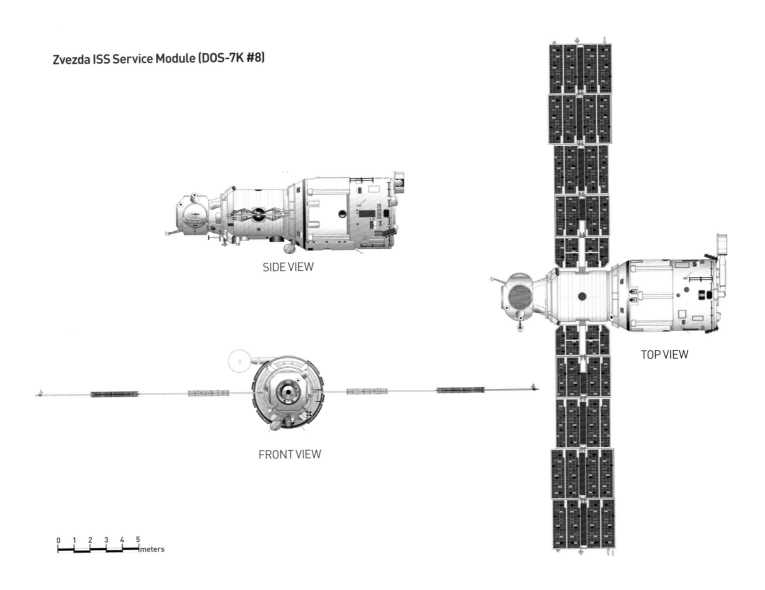

SIDE VIEW

TOP VIEW

FRONT VIEW

0 1 2 3 4 5 meters

on, and during the wait, the shuttle *Discovery* visited the ISS in May 1999 to deliver supplies and logistics, as well as to mount a Russian cargo crane on the outside of the station; and in May 2000 *Atlantis* and its crew boosted the station's orbit in preparation for Zvezda, in addition to installing hand rails on Unity's exterior.

At last, the Khrunichev State Research and Production Center in Moscow completed Zvezda. The engineers at Khrunichev fashioned a spacecraft based on the core component of the Mir space station, which itself derived from the earlier Salyut and Almaz stations. The 39,796-pound (18,051-kilogram), 43-foot (13.1-meter) long Zvezda took off from the Baikonur Cosmodrome in Kazakhstan on July 12, 2000, aboard a Proton rocket. Because the Russians decided to link the ISS and Zvezda by remote control,

some intricate choreography needed to unfold. First, automated computer commands on the spacecraft awakened the onboard equipment and opened its solar panels, measuring 97 feet (29.7 meters) from end to end. Controllers on the ground then oriented Zvezda toward the sun to generate a buildup of electrical power, after which its two main engines fired twice to raise its altitude to that of the ISS. Next, Zarya-Unity approached and rendezvoused with Zvezda on July 25, following which, in a twenty-five-minute sequence, Zvezda's forward port locked onto Zarya.

At this point, the ISS nerve center gradually transitioned from Zarya to Zvezda. Equipped with thirteen windows for observation, Zvezda assumed the infrastructure needs of the early ISS, including electrical distribution (from the solar arrays), data processing (with ESA computers), propulsion (to maintain orbit), communications (data, voice, and television with mission control), life support (recycling wastewater as oxygen), and the management of flight control information. It also served the human needs of its occupants. Its three pressurized holds—in the forward position, a ball-shaped transfer compartment with an airlock; in the center, a cylindrical work compartment; and aft, a spherical transfer chamber—provided sleeping quarters, personal hygiene facilities, a galley with dining tables, and space for a treadmill and stationary bicycle.

With these transfers, Zarya's main function dwindled to storage for external fuel tanks and for other materials. The ISS itself now only awaited its first three-person crew, known as Expedition 1, to begin the long period of human habitation.

Destiny Laboratory

With the joining in July 2000 of the Russian Zvezda ("star") module to the initial Unity and Zarya components of the International Space Station (ISS), the preparatory period ended for the world's first multinational domicile in space. It ushered in a new, inclusive age of human habitation in orbit.

The first crew arrived on November 2, 2000, on which date Mission Commander William Shepherd, Soyuz Commander Yuri Gidzenko, and Flight Engineer Sergei Krikalev boarded Zvezda and inaugurated ISS Expedition 1. After a period of adaptation and adjustment, on the one hundredth day of their journey (February 7, 2001), they awaited the arrival of the shuttle *Atlantis* with a key component of the station tucked into its cargo hold. It docked with the ISS two days later, and for a week the three men and one woman aboard *Atlantis* worked together with the three ISS crewmembers to mate the new member to the forward part of Unity. They conducted three spacewalks, after which *Atlantis* departed, leaving behind a station bigger than any that existed to date: 171 feet (52 meters) long, 90 feet (27 meters) high, and 240 feet (73 meters) wide. Even at this initial stage, it weighed about 224,000 pounds (101,600 kilograms).

The new addition—the American Destiny Laboratory—represented a costly but essential ingredient of the ISS's science mission. Built by the Boeing Company at NASA's Marshall Space Flight Center for $1.4 billion, it looked deceptively unassuming on the outside: a big, silver cylinder the size of a business jet with hatches at both ends. But its simplicity belied its complexity. The gleaming white rectangular interior, divided into four sections, contained six racks apiece, or twenty-four in all. Each of these units measured 73 inches (185 centimeters) tall and 42 inches (106 centimeters) wide. Of the twenty-four, thirteen held science experiments; the remaining eleven serviced the infrastructure of the ISS, producing electrical power, cooling water and

purifying it, adjusting temperature and humidity, and refreshing the atmosphere.

The science payloads on Destiny varied from rack to rack. Researchers from around the world relied on such items as electrical and fluid connectors, sensors, video cameras, motion dampeners, and other pieces of equipment to carry out their experiments by remote control. Most of the work shared a common objective: to assess the long-term effects of zero gravity on physical and biological processes, and particularly on human beings and their environment. In pursuit of this knowledge, disciplines such as ecology, earth science, chemistry, biology, physics, and biomedicine shared a stake in Destiny's experiments.

Destiny also served Earth-based meteorologists and geologists with an optical-quality window, through which crew members photographed some majestic images of the Earth's topography, weather systems, and natural occurrences such as fires, floods, and avalanches.

Multipurpose Logistics Modules

If the Destiny Laboratory inaugurated science research aboard the International Space Station (ISS), the arrival of the multipurpose logistics modules (MPLMs) heralded a new phase in ISS operating efficiency. Designed and fabricated in Italy by Thales Alenia Aerospazio for the Italian Space Agency (ASI), a total of three MPLMs left the factory in Turin. All bore the names of renowned artists of the Italian Renaissance: Leonardo, in honor of the polymath painter, sculptor, and architect Leonardo da Vinci (1452–1519); Raffaello, in tribute to the architect and artist Raffaello Sanzio (1483–1520); and Donatello, in recognition of the sculptor Donato di Niccolo Di Betto Bardi (1386–1466). They arrived at the Kennedy Space Center (KSC) aboard the Airbus Beluga transport in August 1998, August 1999, and February 2001, respectively.

In all, the MPLMs visited the shuttles twelve times, from March 2001 to July 2011, with a substantial gap between 2003 and 2005 due to the loss of the shuttle *Columbia*. Leonardo went into orbit eight times: in March and August 2001, June 2002, July 2006, November 2008, August 2009, April 2010, and finally in February 2011—always in the payload bay of either *Discovery* or *Endeavour*. In preparation for its last mission, Leonardo underwent extensive modifications at KSC with parts (particularly the multilayer insulation blankets) cannibalized from Donatello, which never ventured into space. When Leonardo returned to the station in February 2011, it stayed there as a fixed part of the ISS (redesignated as the permanent

US Lab "Destiny"

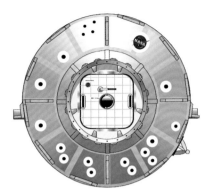

FRONT VIEW

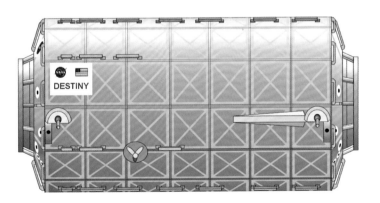

TOP VIEW

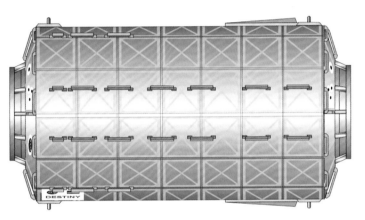

SIDE VIEW

0 1 2 3 4 5
meters

MPLM *Raffaello*

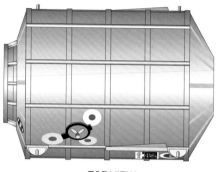

TOP VIEW

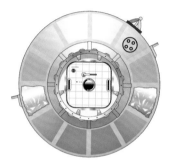

FRONT VIEW

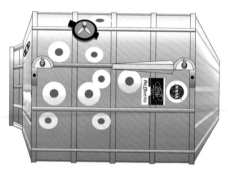

SIDE VIEW

0   1   2   3   4   5
meters

multipurpose module, or PMM). Raffaello traveled to the station four times: in April and December 2001, July 2005, and July 2011, on *Discovery*, *Endeavour*, and *Atlantis*.

Some refer to the MPLMs almost dismissively as space vans or space trucks, but these terms underestimate their value. These 21-foot-long (6.4-meter), 15-foot-wide (4.6-meter), 9,000-pound (4,100-kilogram) pressurized cylinders carried cargo weighing up to 20,000 pounds, arranged on a system of sixteen racks. Upon the arrival of an MPLM at the ISS, the cosmonauts and astronauts unloaded supplies, spare parts, equipment, and new experiments. Before they returned to Earth (on board one of the shuttle orbiters), the crews packed them with completed research projects and with refuse.

But in addition to their logistics functions, during the times that Leonardo and Raffaello docked with the station they acted as fully functioning ISS components, providing the life support, fire suppression, electrical distribution, and computing power required to create a livable environment for its human occupants. Further, the MPLMs served as the foundational pattern for two other major ISS modules. Harmony, built for NASA by Thales Alenia Aerospazio in Italy, arrived in October 2007 and served as

both the second ISS node (after Unity, Node 1), and as the new American crew quarters. And the European Space Agency's Columbus laboratory, launched in February 2008, also owed a design debt to the MPLMs.

Tiangong-1

Unlike the Shenzhou capsule that owed such a debt to the Russian space agency and to Soyuz, the development of a large modular space station gave the Chinese National Space Agency (CNSA) the opportunity to demonstrate its independence and sophistication. The CNSA set about this task not only because it could (due to its technical ability), but also because of misgivings about its exclusion from the International Space Station (ISS), as well as restrictions against its collaboration with NASA (imposed by congressional action in 2011). Perhaps CNSA's leaders interpreted the lack of options as a go-ahead to proceed with their own program.

Although exploratory discussions about a station began in China as early as 1992, official authorization came in 1999—perhaps not coincidentally, the year after the first two parts of the ISS (the Unity and Zarya modules) mated in space. Chinese

CAST Tiangong 1 (Project 921-2)

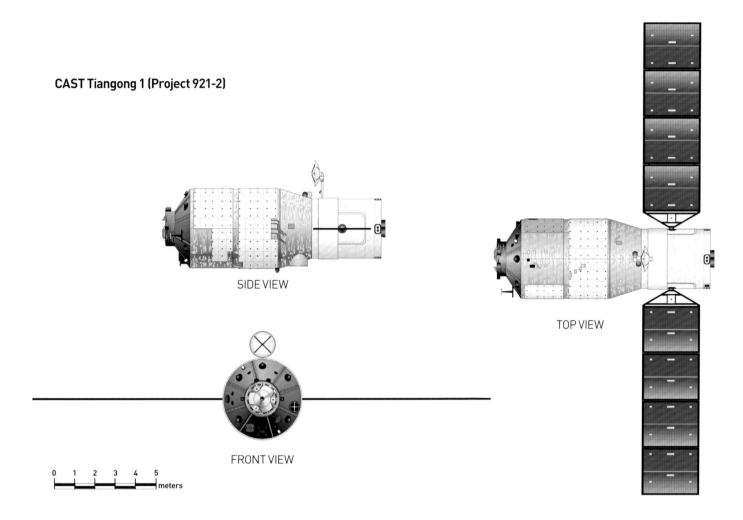

SIDE VIEW

TOP VIEW

FRONT VIEW

0 1 2 3 4 5
meters

authorities decided to build three small testbeds, leading to a highly ambitious goal: an outsized station with a 20-ton core buttressed by two smaller research modules and a cargo spacecraft. CNSA scheduled its completion between 2020 and 2023—a timeframe coincident with the ISS's retirement.

During 2011 and 2016, the Chinese space agency sent two of the preliminary stations into orbit. The first, called Tiangong (Heavenly Palace)-1, served dual purposes: as a science laboratory and as a target vehicle for manned and autonomous docking by the Shenzhou capsule. Although intended only as a step toward the full-sized station of the future, Tiangong-1— measuring 34.1 feet (10.4 meters) long, 11 feet (3.35 meters) in diameter, and weighing 18,753 pounds (8,506 kilograms)— represented a historic achievement in itself. Launched on September 29, 2011, it consisted of two cylindrical sections: at the forward end, a wider, habitable experimental module with 530 cubic feet (15 cubic meters) of interior space; and aft, a narrower resource module, on which engineers mounted electrical systems, environmental controls, two solar wings (spanning about 75.5 feet or 23 meters), and the station's propulsion apparatus. The experimental module provided two

beds and exercise equipment for the Chinese astronauts (called taikonauts). The Shenzhou capsule—which carried the crew and remained in place for the duration of each mission—augmented the experimental module with a third bed, cooking facilities, and a toilet. Tiangong-1's design incorporated berths at both ends of the structure, but only the one on the experimental module accommodated automated docking.

Tiangong-1 hosted two manned (Shenzhou 9 and 10) visits, and one unmanned (Shenzhou 8) before it ended service in March 2016, following a two-year extension of its career. Subsequently, in September 2016, CNSA announced that it lost control of the spacecraft and that it reentered and burned up on April 2, 2018. Nonetheless, the program continued: Tiangong-2 went into orbit on September 15, 2016, but instead of flying Tiangong-3 later as planned, the Chinese leaders decided to merge it with the Tiangong-2 mission.

So, despite its status as a relative newcomer to space exploration, CNSA showed a drive and a record of accomplishment that, if borne out by the success of its upcoming master station, promised to close much of the gap between it and NASA, the Russian, and the European space agencies.

ROCKETS
Space Launch System

When President George W. Bush decided to terminate the space shuttle in 2010, NASA faced uncertainty about how to transport astronauts into space. Bush offered the solution in 2004 in the form of a sweeping program ultimately called Constellation. Constellation depended on two rockets and a capsule: the Ares I launch vehicle to lift a new, Apollo-like spacecraft called Orion into orbit; and the far larger Ares V to bring supplies and equipment to Orion prior to its voyages to the moon, and later to Mars. Although it seemed plausible because it borrowed heavily from the technologies of Apollo and the space shuttle, Constellation came to a halt in April 2010 when President Barack Obama announced its cancellation (except for Orion) in a speech at Kennedy Space Center (KSC). In its place, he pledged to increase NASA's budget by $6 billion and directed the space agency to develop a massive launch system that looked to the future, rather than the past, for its technical inspiration.

Despite its differences from the Constellation vehicles, the big rocket pursued by NASA under the Obama administration did not ultimately represent a clean break with tradition, regardless of the president's wishes. Even with more funding, NASA still lacked the resources to redefine the basic tenants of chemical rocketry. So after about a year and a half of design work, in September 2011 the space agency revealed the architecture of a new, nonreusable booster known as the space launch system (SLS).

Rather than building two separate rockets, as proposed under Constellation, the SLS involved just one, but was based on a progressive design philosophy that expanded in capacity over time. NASA plans called for the initial SLS to carry 154,000 pounds (70 metric tons) into orbit, but in its final form to almost double the payload to about 286,000 pounds (130 metric tons). President Obama may have called for imaginative new designs and technologies in his KSC speech, but NASA Deputy Administrator Lori Garver expressed the budgetary realities faced by her agency when she said that NASA sought to realize Obama's objectives, while also being mindful of the cost.

Practically, this meant doing just as Constellation did: to the maximum extent possible, using existing, off-the-shelf components and modifying them as required for the new project. The SLS Block 1—322 feet (98.1 meters) long and 28 feet (8.4 meters) in diameter (compared to the Saturn V at 363 feet (111 meters) long and 33 feet (10 meters) in diameter)—consisted of four main parts. The biggest, a 200-foot-long core stage reminiscent of the space shuttle external tank, held 730,000 gallons of liquid hydrogen and liquid oxygen, the propellant for four modified Aerojet/Rocketdyne RS-25 shuttle main engines mounted at the bottom of the stack and capable of delivering over 2 million pounds of thrust collectively. Attached to the sides of the core stage, two shuttle-derived solid rocket boosters each provided 3.6 million pounds of thrust from their QM-1 motors. Finally, the interim cryogenic propulsion stage (ICPS)—better known as the second stage of the Delta IV missile—consisted of a single RL-10 engine fueled by liquid hydrogen and liquid oxygen that traced its lineage to the venerable Centaur rocket. It produced 24,750 pounds of thrust.

Work on the SLS progressed according to a series of milestones. In December 2011 (three months after NASA released its technical objectives), the space agency signed contracts with Boeing and others to transfer as much as of the Ares project as possible to SLS. A year later Boeing's core stage proposal cleared a preliminary design review. In the end, NASA and Boeing agreed to a prime contract in July 2014 for $2.8 billion, extending to 2021. Finally, the SLS passed a critical design review in October 2015—a turning point at which experts inside and outside of NASA declared the program ready to proceed with manufacturing, integration, and testing.

Boeing integrated the parts of the core stage at NASA's Michoud Assembly Facility near New Orleans, Louisiana, and technicians finished the welding on its liquid fuel tank in September 2016. In January 2017, the NASA Marshall Space Flight Center in Huntsville, Alabama, completed a two-year construction project consisting of two massive towers—a structural test stand on which space agency personnel planned to assess the integrity of the SLS core stage fuel tank. Following these steps, NASA scheduled a barge ride for the core stage from Michaud to NASA's Stennis Space Flight Center in Hancock County, Mississippi, for hot fire testing.

Initially, NASA's SLS timetable called for the first unmanned flight (a voyage around the moon) in 2017; it later was delayed until November 2018. The space agency anticipated a lunar circumnavigation with four astronauts on board in August 2021.

Even though its designers based SLS on modified rocket and capsule technologies from the Apollo and shuttle eras, America's first deep space initiative since Saturn V cost upward of $7.75 billion; and that only in the 2011 to 2015 timeframe.

Falcon 9

When President Barack Obama announced at the Kennedy Space Center (KSC) in April 2010 the commercialization of orbital spaceflight, he set in motion the construction not only of new spacecraft, but of new rockets as well.

Space Launch System (SLS Block 1)

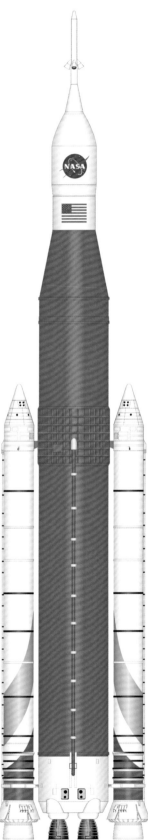

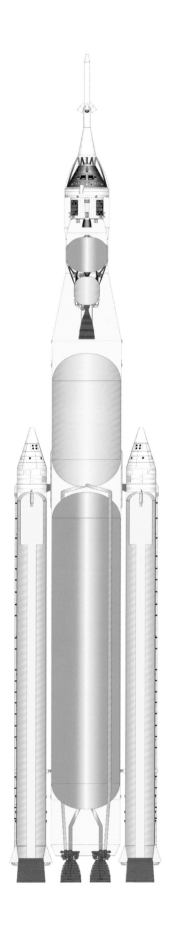

Space X Falcon 9 Full Thrust

0 — 5 meters

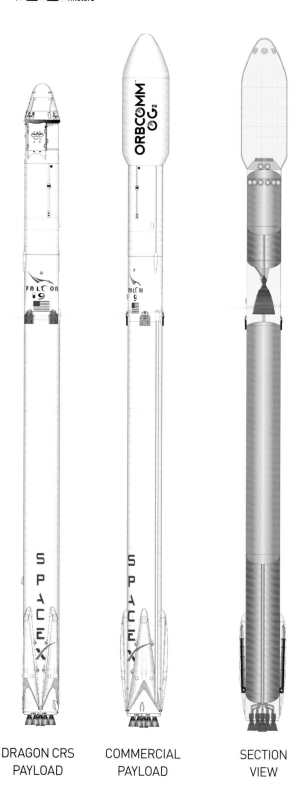

DRAGON CRS
PAYLOAD

COMMERCIAL
PAYLOAD

SECTION
VIEW

SpaceX, of Hawthorne, California, which opened its doors eight years before the president's speech, had an early start among its industrial competitors. It funded the development of its first booster—the Falcon 1—from its own resources and made five flights from 2006 to 2009. Meanwhile, in 2006, the company won seed money from NASA's Commercial Orbital Transportation Services (COTS) program for three preliminary launches of a bigger rocket then under construction, Falcon 9. Two years later, NASA purchased twelve Falcon 9 resupply missions to the International Space Station (ISS), pending successful demonstration flights. Aided by good timing, Falcon 9 made its first launch in June 2010—just a month after President Obama initiated the commercial launch venture in his KSC speech.

Like the Dragon capsule, the two-stage Falcon 9 emphasized reliable operation. SpaceX's designers accomplished the separation of stage 1 from 2 not with the explosive charges common to most launch vehicles, but with a far safer, low-impact pneumatic system. And the engineering team added redundancy to stage 1: a ring of nine Merlin engines fired at liftoff, with only seven actually needed to achieve orbit. The first stage tankage—constructed from high-strength aluminum-lithium alloy—supplied the liquid oxygen and kerosene mixture that powered the engines. Stage two, fashioned from the same materials, contained a single Merlin that sent the payload into orbit; or many payloads into different orbits because of the engine's ability to start repeatedly.

In all, the two stages produced 1.7 million pounds of thrust at sea level. The total stack measured 230 feet (70 meters) long and 12 feet (3.7 meters) in diameter. It weighed 1,208,000 pounds (549,000 kilograms) and lifted a payload of up to 50,265 pounds (22,800 kilograms).

During its first seven years of operation (2010 to 2017), Falcon 9 experienced a wide variety of launches, with thirty-seven successful flights in all during this period (to June 2017). It flew one test launch, made twelve flights with the Dragon capsule, and lifted twenty-five satellites for a variety of customers, from the US Air Force to NASA (the Jason-3 satellite), and from Thailand's Thiacom Public Company to the Italian Thales Alenia Space Corporation. Of these thirty-seven, two minor (but highly noticed) failures occurred: after launching their cargo, two Falcon 9 rockets missed in attempts in January and April 2015 to land on barges in the Atlantic Ocean—a critical demonstration of reuse. Subsequently, two major failures occurred in June 2015 and September 2016, when Falcon 9 rockets exploded near or on the launch pad. In between these two disasters, on December 22, 2015, SpaceX achieved a milestone denied earlier; after sending into orbit a satellite for Orbcomm, the Falcon 9 flew back to Earth and landed vertically on a vessel floating in the Atlantic.

Following the September 2016 explosion, the company stopped launch operations temporarily and made technical assessments. When it returned to flight on January 14, 2017, SpaceX made headlines again: the Falcon 9 booster once more touched down on a platform, this time in the Pacific Ocean, after launching the Iridium 1 satellite.

Subsidized in part by NASA contracts for ISS resupply and human flights, SpaceX achieved a great deal in a short time, giving credence to the Obama administration's decision to contract orbital space services from the private sector. But SpaceX also set its sights on the next generation of rocketry, the Falcon Heavy, potentially the most powerful US rocket since Apollo's Saturn V. As conceived, the Falcon Heavy consists of a Falcon 9 core and two strap-on boosters, capable of lifting 120,000 pounds (54,431 kilograms)—more than twice the payload of Falcon 9.

SpaceX anticipated a first launch of Falcon Heavy in November 2017, later delayed until January 2018. But as it prepared for this historic event, it again became embroiled in a mishap, although probably not of its own making. On Sunday, January 7, 2018, a Falcon 9 lifted off from Cape Canaveral

Space X Falcon 9 Full Thrust

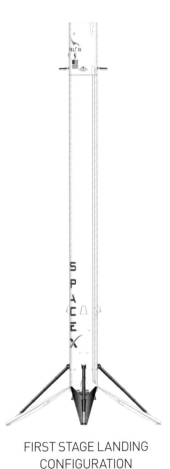

FIRST STAGE LANDING
CONFIGURATION

FIRST STAGE LANDING
CONFIGURATION
(45° ROTATED)

Space X Falcon Heavy

Space X Falcon Family

0 1 2 3 4 5 meters

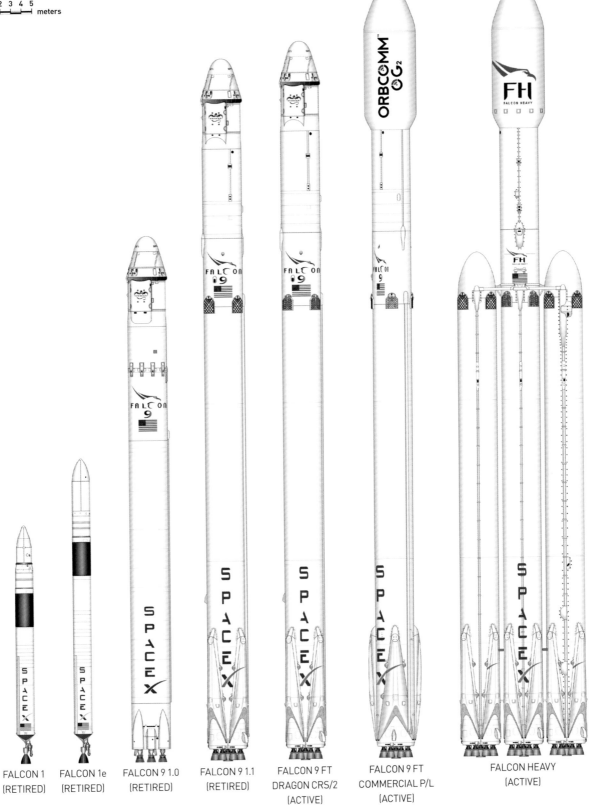

FALCON 1
(RETIRED)

FALCON 1e
(RETIRED)

FALCON 9 1.0
(RETIRED)

FALCON 9 1.1
(RETIRED)

FALCON 9 FT
DRAGON CRS/2
(ACTIVE)

FALCON 9 FT
COMMERCIAL P/L
(ACTIVE)

FALCON HEAVY
(ACTIVE)

carrying a top-secret satellite named Zuma, thought to be a multibillion-dollar spacecraft. The flight appeared to go normally as the rocket's second stage and payload entered orbit. But soon, stories circulated that the mission had gone awry. On January 9, *The Wall Street Journal* cited industry and government officials who believed that Zuma failed to separate successfully from the Falcon 9's second stage, and that the two fell back into the atmosphere for a total loss.

SpaceX's president Gwynne Shotwell issued a terse statement saying that their rocket did "everything correctly" and that a review of the launch data showed that "no design, operational, or other changes are needed." She added that "information published that is contrary to this statement is categorically false." Northrop Grumman, which fabricated the Zuma satellite and perhaps the adapter to the Falcon's second stage, offered no explanation due to the classified nature of the project. By the time of the publication of *The Wall Street Journal* article, House and Senate staffs had been briefed on the incident.

SpaceX's unequivocal denial of culpability, bolstered by the company's uninterrupted launch schedule, offered reassurance to its growing list of customers. This air of confidence proved to be well founded when, on February 6, 2018, SpaceX launched the maiden flight of the Falcon Heavy from the Kennedy Space Center's historic pad 39A—the spaceport for the Saturn V rocket and the space shuttle. The mission not only succeeded, but the rocket's two reusable side boosters landed almost simultaneously on site at Cape Canaveral. The center booster, however, missed its landing point on a barge floating in the Atlantic. Still, this event heralded more than a technical achievement; for the first time in the space age a private firm, rather than a governmental agency, developed, built, and sent into space a launch vehicle powerful enough to send human beings and other cargo beyond Earth's orbit.

Antares

Among the smaller companies vying for a share of NASA's commercial launch service contracts, the longest record of achievement belonged to Orbital ATK. Founded at Dulles Virginia in 1982, the firm (then known as Orbital Sciences Corporation, since merged with Alliant Techsystems, or ATK, in 2014) distinguished itself for its small satellite booster services. Orbital adapted an L-1011 airliner and used it as a platform for its three-stage Pegasus solid fuel rocket, the first privately built space launch vehicle. Inexpensive and highly reliable, Pegasus continues to carry many of NASA's most sensitive satellites into orbit. Orbital entered the International Space Station (ISS) resupply competition in 2013 under a commercial orbital transportation services (COTS) contract with NASA for $1.9 billion. It committed the company to eight flights to the station,

Orbital ATK Antares-200

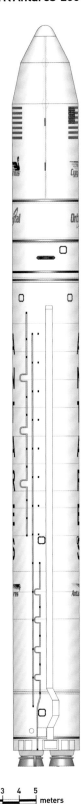

0 1 2 3 4 5
meters

flown by its Cygnus spacecraft and lifted by its Antares rocket. Antares (whose name derived from the giant red star in the constellation Scorpius) consisted of a novel pairing of Eastern and Western design, integrated by Orbital: a liquid propellant (liquid oxygen/kerosene) first stage built by KB Yuzhnoye/Yuzhmash in the Ukraine; two RD-181 first stage engines made by NPO Energomash, a subsidiary of the Russian firm Energia; and a solid propellant (hydroxyl-terminated polybutadiene) second stage fabricated by Orbital itself. Antares measured 133 feet (40.5 meters) long, almost 13 feet (3.9 meters) in diameter, with a launch mass of between 621,704 and 652,568 pounds (282,000 and 296,000 kilograms). Its nearly 13-foot-(3.9-meter-) long fairing accommodated large payloads.

Cygnus (a northern constellation on the plane of the Milky Way) followed the multinational example of Antares. Its barrellike structure measured 16.7 feet (5.1 meters) long by 10.1 feet (3 meters) in diameter in the standard version; 20.7 feet (6.3 meters) long by 10.1 feet (3 meters) in diameter in the enhanced. Thales Alenia of Turin, Italy, fabricated its pressurized cargo module; Orbital made its service module.

All Antares flights to date launched from NASA's Wallops Flight Facility in Virginia. The initial two (in April and September 2013) proved the flight worthiness of the Antares/Cygnus combination, the first with a simulated Cygnus payload, the second as a NASA COTS demonstrator. Then the Antares began to make automated resupply flights to the ISS, beginning in January 2014 and repeated in July. But disaster struck on October 28, 2014. On this flight, as in the earlier ones, the first stage engines consisted not of the RD-181s that Orbital later selected, but of older Russian NK-33 powerplants retrofitted by Aerojet Rocketdyne as the AJ-26. A NASA investigation of the incident—a massive explosion just after launch—found that about fifteen seconds after ignition one of the liquid oxygen turbopumps in the AJ-26 exploded, the result of rotating and stationary parts coming into contact. After more than a year-long hiatus, in December 2015 Orbital ATK launched another Cygnus, but this time contracting with United Launch Alliance for an Atlas V rocket, which carried 7,700 pounds 3,500 kilograms) of cargo to the ISS. In March 2016, the same vehicle boosted nearly 7,500 pounds (3,400 kilograms) to the station. Antares returned to flight with an upgraded version on October 17, 2016, and successfully sent Cygnus on a resupply mission. NASA anticipated another Antares mission in 2017, but not before a third Atlas V flight in April of that year.

Like SpaceX, Orbital ATK proved itself capable of bringing provisions and equipment to the ISS; but so far, these smaller firms seemed no less immune to setbacks and disasters than the more established firms—or than NASA itself.

New Shepard

Among the firms competing for a share of the commercial space business initiated by President Barack Obama in 2010, Blue Origin LLC—begun by Amazon.com founder Jeff Bezos—distinguished itself by starting relatively late but having one of the most ambitious programs. Blue Origin also entered the contest on a somewhat different footing than the others. Rather than vie for a share of the International Space Station (ISS) resupply and astronaut transportation business, Blue Origin joined with NASA in 2016 as one of six companies competing under the space agency's Flight Opportunities program, devoted to nurturing suborbital missions. This initiative enabled NASA to encourage a variety of providers to offer launch services for promising technologies developed by government, academia, and industry. Blue Origin fielded the New Shepard rocket as its suborbital candidate. The company fabricated its spacecraft in Van Horn, Texas, and launched and landed them near this site. Early in its evolution, Blue Origin relied on several sources of revenue: NASA, researchers willing to pay for access to space, and private citizens ready to sign up for a future astronaut experience.

Like Falcon 9, New Shepard represented an attempt at using the booster repeatedly, rather than once, and differed from Falcon 9 in that it landed on solid ground, rather than at sea. Blue Origin's reuse protocol began after a 2.5-minute launch sequence by New Shepard, at which point (in order of events), its engines cut off, the booster fell to Earth, its drag brakes engaged, its rockets fired, and it deployed landing gear for a vertical touch down. Before New Shepard began its descent, Blue Origin's passenger capsule separated from the rocket and coasted into space before returning to Earth on parachutes.

Blue Origin designed, fabricated, and tested its engines at its headquarters in Washington state. It began small with the kerosene propellant Blue Engine (BE)-1, capable of 2,200 pounds of thrust at sea level. Then it built the BE-2, fueled by kerosene and peroxide for 31,000 pounds of thrust; developed the BE-3 (New Shepard's engine), a liquid hydrogen/liquid oxygen mixture capable of 110,000 pounds of thrust; and planned for the BE-4, powered by liquefied natural gas/liquid oxygen fuel and designed to produce 550,000 pounds of thrust. Blue Origin expected eventually to scale up the New Shepard to a two- and three-stage heavy booster, respectively 270 feet (82.3 meters) and 313 feet (95.4 meters) long; equip it with seven BE-4 engines; and under the name New Glenn, use it for orbital missions. The capsule for New Glenn will seat six astronauts in 530 cubic feet of space.

During 2015 and 2016, New Shepard made five full-scale test flights. The first, in April 2015, lifted an unmanned crew

Blue Origin New Shepard

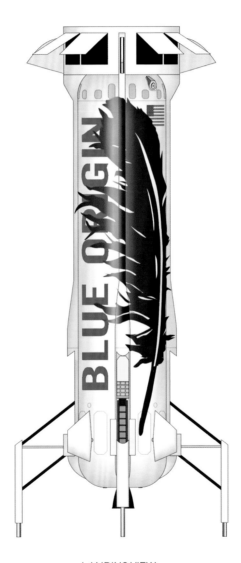

SIDE VIEW

LANDING VIEW

SIDE SECTION

0 1 2 3 4 5
meters

Arianespace Ariane 5 GS
(V 193 18 December 2009)

0 1 2 3 4 5 meters

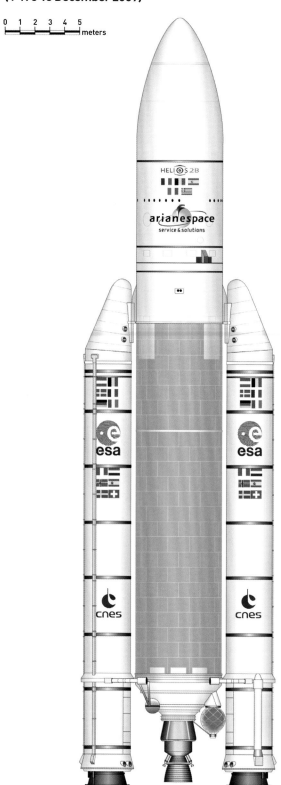
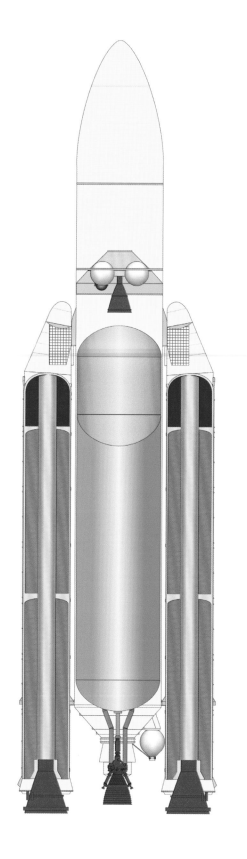

capsule to 307,000 feet (93,574 meters), but the booster could not be recovered as planned due to a loss of pressure in the hydraulic system. The next flight in November offered greater success, sending the capsule to almost 330,000 feet (100,584 meters) and, as reported by media around the world, ending with New Shepard making a vertical landing on West Texas soil—the first achieved anywhere. New Shepard flew again in April 2016, testing a new reaction control system, conducting two microgravity experiments, and again recovering the booster. The fourth flight in June 2016 successfully tested a double, rather than the standard triple, string of parachutes for the capsule descent and once more demonstrated a safe touch down for New Shepard. Finally, in October 2016 the rocket's crew abort system underwent a test at an altitude of 16,000 feet (4877 meters). The capsule separated as planned, and to the surprise of mission planners, New Shepard again made a successful vertical landing—for the fourth consecutive time. Significantly for the future of reusable systems, the same, refurbished vehicle landed without incident in each mission.

Although not yet a competitor for the lucrative ISS resupply business, Blue Origin showed a capacity for innovation in its chosen part of the emerging commercial space launch market. The company anticipated its first crewed flights in 2018.

Ariane 5

Despite the European Space Agency's (ESA's) extraordinary expansion of launch capability in less than a decade (1979 to 1988), the biggest leap began during the middle of that period. ESA approved the development of the French-made, heavy-lift Ariane 5 in January 1985, and with it inaugurated a program that gave Europe a capability competitive with that of any of the world's space powers. Compared to its Ariane 4 predecessor, it made huge gains, doubling the thrust and more than doubling the weight of its payloads.

Added to that, ESA flew its rockets from a spaceport that rivaled that of its competitors. In 1964, the French government selected a location in one of its overseas departments—in Kourou, French Guiana—as the site of its launches. When ESA came into being in 1975, the French authorities proposed sharing this complex with the new agency. Since that time, ESA has paid two-thirds of Kourou's annual budget and in addition covered the costs of facilities construction. Not just European, but also the American, Indian, Canadian, Brazilian, and Japanese governments and commercial firms rely on this spaceport for orbital flight.

Building the space center in French Guiana proved to be a wise choice. Located only 311 miles (500 kilometers) north of the equator, its geographical position enabled launches of geostationary satellites with few changes to their trajectory and offered a slingshot effect that boosted the speed of rockets by about a quarter of a mile (460 meters) per second. It also offered conditions favorable to safe operations: a small local population not likely to encroach on the facility; surrounding terrain covered almost entirely by equatorial forests; and rare disruptions from earthquakes or cyclones. But one big hurdle did exist: the vast distance between the manufacturing plants in Europe and the launch pads. The compromise solution lay in shipping the enormous cargo in pieces across the Atlantic, with the only disadvantages being transit time (nine weeks for portage, unpacking, and setup prior to liftoff) and associated expenses.

Ariane 5 made its maiden flight from Kourou in June 1996. Its French designers, like their Russian counterparts, preferred to modify, rather than terminate older spacecraft and to keep them in service for long periods. Accordingly, the Europeans produced five Ariane 5 varieties over twenty years (the G, G+, GS, ECA, and the ES) and signed contracts with Arianespace—a multinational consortium that handled the commercial aspects of Kourou's activities—to continue Ariane 5 launches until 2023.

Although it had an expendable system like the earlier Arianes, Ariane 5's role as a heavy-lift launch vehicle distinguished it from its other family members. The Ariane 5 ranged in height (depending on type) from 151 to 171 feet (46 to 52 meters), measured about 18 feet (5.4 meters) in diameter, and weighed roughly 1,713,000 pounds (777,000 kilograms) at liftoff. The ECA model carried a 23,100-pound (10,500-kilogram) payload into geosynchronous transfer orbit, and the ES, 44,000 pounds (16,000 kilograms) into low Earth orbit.

Ariane 5 ECA derived its power from three principal components. Its cryogenic main stage consisted of a Vulcain 2 engine situated at the base, fueled by propellants pumped from a 100-foot (30.5-meter) tank divided into separate liquid oxygen and liquid hydrogen compartments. It generated about 300,000 pounds of thrust. Attached to the sides of the main stage, two solid rocket boosters fed by ammonium perchlorate, aluminum fuel, and polybutadiene delivered a combined 2,600,000 pounds of thrust. The second stage, known as the cryogenic upper stage, relied on one HM7B engine, powered by liquid oxygen and liquid hydrogen, which added 14,000 pounds of thrust to the rocket's combined total of over 2,900,000.

Ariane 5 compiled an enviable launch record. From its first flight in June 1996 to one twenty-one years later—in June 2017—it flew ninety-four times with only two outright failures (in 1996 and again in 2002) and two partial failures (in 1997 and 2001). Over its long lifespan, it has carried an enormous variety of spacecraft, such as ESA's automated transfer vehicle (an expendable cargo spacecraft for the International Space Station, XMM-Newton, Rosetta, and the Herschel-Planck Space Observatory), the European Union's Galileo global navigation system, and DirecTV and Intelsat commercial satellites. It will also lift the famed James Webb Space Telescope into space, scheduled for May 2020.

Despite the difficulties inherent in winning consensus and cooperation among the ESA Governing Council's twenty-two member states, the exceptional record of the Ariane family of rockets—one that spans almost forty years of launch history—proves that spaceflight need not be pursued exclusively on the basis of national rivalries or of national pride.

VEGA

Compared to the step-by-step, calibrated progression of the European Space Agency's Ariane family of rockets, the initiation of the VEGA booster seemed almost off-handed. The project came into focus in 1988, the same year Ariane 4 made its first flight. During that year, the United States retired a Scout launcher that serviced small Italian satellites from a floating platform off the Kenyan coast. To fill the void, an Italian firm—BPD Difesa Spazio—proposed to the newly created Italian Space Agency (ASI) the fabrication of a small, home-grown rocket using the Zefiro motor developed for the Ariane program. Then, during the early 1990s, several Italian aerospace industries suggested pairing the solid booster technology that they contributed to Ariane 4 and 5 with the Zefiro. They called the project the Vettore Europeo di Generazione Avanzata (Advanced Generation European Carrier Rocket)—known by the Italian acronym VEGA. Not coincidentally, it was also the name of the brightest star in the constellation Lyra. By April 1998 ESA's governing council initiated the VEGA's predevelopment phase, basing it on a configuration that enabled its first stage to double as an improved Ariane 5 strap-on rocket.

VEGA got the full ESA go-ahead in November 2000, and in the following month Italy, France, Switzerland, Sweden, Spain, the Netherlands, and Belgium agreed to finance the project. These contributors and ESA pursued VEGA with a straightforward goal: to field a low-cost, lightweight vehicle with high reliability, primarily for scientific and Earth-observation satellites at polar and sun-synchronous orbits. Italy took the lead in this international consortium; the Italian Space Agency and the Fiat Avio corporation joined forces as prime contractor. Arianspace assumed its standard role, taking charge of VEGA mission management and launch services at ESA's Kourou flight facility in French Guiana.

VEGA underwent a long development and many postponements. It was originally scheduled for its first launch in 2006, but program managers pushed the launch back until 2007. And then the date was moved to 2009, in part due to construction at Kourou to convert the ELA-1 launch pad (used for the Ariane 1) into one suitable for VEGA. However, not until 2011 did the new rocket arrive in French Guiana for the initial mission. At last, on February 13, 2012, VEGA launched its first payload: an 860-pound (390-kilogram) Laser Relativity Satellite (LARES) for the Italian Space Agency.

At the same time, it also sent into orbit the 27.5-pound (12.5-kilogram) Alma Mater Satellite (ALMASat-1)—a technology demonstrator fabricated at the University of Bologna—as well as a constellation of seven tiny spacecraft called CubeSat. A mere 4 inches (10 centimeters) wide and weighing no more than 2.2 pounds (1 kilogram) each, the Cubesat constellation served as a trainer for engineering students studying satellite development and operations.

VEGA gave substance to the expression that less can be more. Just 98 feet (30 meters) tall and 9.8 feet (3 meters) in diameter, it weighed only 302,000 pounds (137,000 kilograms)—smaller even than the dimensions of Ariane 1 (164 feet [50 meters] in height, 12.4 feet [3.8 meters] in diameter, with a mass of 456,700 pounds [207,200 kilograms]). But VEGA proved its value. Its ability to lift several small satellites into orbit at once reduced launch costs for Arianspace's customers. And despite its size, VEGA actually carried almost as much payload weight compared to the bigger and heavier Ariane 1. VEGA's first three stages (powered by a P80 engine in the first, Zefiro 23 in the second, and Zefiro 9 in the third) flew on solid propellant. Only the stage four RD-843 powerplant used liquid propellant (unsymmetrical dimethylhydrazine and nitrogen tetroxide).

During its first five years of flight, VEGA established a perfect record, with nine successful flights. Six of these missions involved Earth observation satellites, but it also sent ESA's LISA Pathfinder into orbit (see the profile of LISA Pathfinder on page 216) and conducted reconnaissance and suborbital missions, satisfying clients as diverse as the governments of Peru, Turkey, and Vietnam. Six more flights remained on the manifest up to 2020.

As a complement to its bigger Ariane rockets, ESA and Arianspace recognized a niche market for multiple payload, small-scale commercial launches, and filled it ably with VEGA.

Long March-2F

In one of the most consequential ironies of the Cold War, the modern Chinese ballistic missile and rocket program began not in Beijing or Shanghai, nor at a remote launch site, but in the United States, in the quiet suburb of Pasadena, California, at the California Institute of Technology (Caltech).

Avio–VEGA Launcher

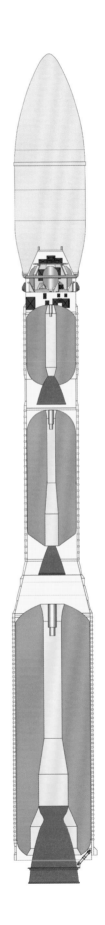

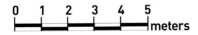

0 1 2 3 4 5
meters

CAST CZ-2F

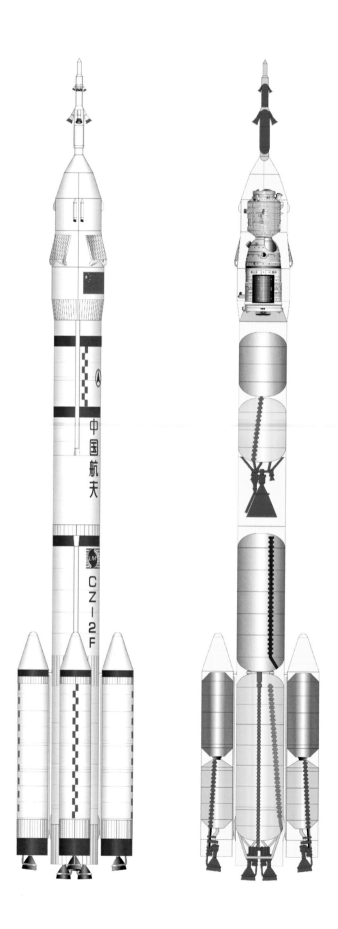

0 1 2 3 4 5 meters

There, in 1936, the famous professor of aeronautics, Hungarian-American Theodore von Kármán (1881–1963), welcomed a Chinese student named H.S. Tsien (1914–1997) to do advanced study toward his doctorate. Tsien and a small cohort of other students hoped to study rocketry, and partly because Kármán considered Tsien one of his brightest students, he expanded the Caltech program to accommodate them. Tsien, Frank Malina, and the others eventually built test stands in the nearby Arroyo Seco canyon, which in the 1940s became the grounds of the Jet Propulsion Laboratory. In time, Kármán came to regard Tsien as a trusted colleague, described him as a genius, and even asked him to join an official delegation sent to Europe at the end of World War II to survey cutting-edge aeronautical technologies for the US Army Air Forces.

Tsien graduated in 1947, taught briefly at MIT, and returned to Caltech in 1949—a year of great turbulence for him and the United States. Not only did the communists rise to power in China, but in America the fear of communist influence manifested itself in the pursuit of individuals—whether guilty or not—suspected of communist sympathies. Tsien fell under suspicion when he refused to testify in a perjury case of a coworker with alleged communist ties. The FBI took notice in part because of Tsien's nationality, and in part because he had been working on classified rocket research for a number of years.

The situation only worsened for him. Officials revoked his security clearance, and he threatened to return to China if they did not reinstate it. At this point, he underwent two weeks of detention, and despite the absence of evidence, he found himself prohibited by immigration authorities from leaving the United States for five years. Embittered and at times deeply depressed, Tsien finally left for home with his wife and children after his deportation in 1955.

When he arrived, government authorities asked him to lead the Chinese ballistic missile program. The following year, he assumed a broader and more formal role as director of the Fifth Academy of National Defense, charged with ballistic missile and atomic bomb development. Then, in 1965, his organization began research on a new program, the Long March-1 (Changzheng-1) rocket, the first Chinese satellite launcher. Its two lower stages derived from the Dong Feng-3 intermediate range ballistic missile and its third was from a newly designed solid rocket motor. It made its initial flight—an unsuccessful one—on November 16, 1969. But on April 24, 1970, China joined the small club of nations with an orbital launch capability when the Long March-1 lifted a navigation satellite into space.

Meanwhile, Tsien and his lieutenants set their sights on an even higher target. During the 1970s, they began to develop a more powerful rocket, Long March-2, based on the Chinese DF-5 ICBM. The series began with the modest Long March-2A, a two-stage vehicle almost 105 feet (32 meters) long with a liftoff weight of 380,000 pounds (172,365 kilograms). By the 1990s, the much more powerful Long March-2F appeared on the drafting tables at the Chinese Academy of Launch Vehicle Technology (CALT), the country's main rocket manufacturer. A giant next to the -2A, it measured 203 feet (62 meters) in height, 11 feet (3.35 meters) in diameter, with a mass of 1,023,000 pounds (464,000 kilograms). It consisted of a stage one powered by 4 YF-20B engines, a second stage with 1 YF-24B engine, and four YF-20B strap-on boosters.

Its purpose became clear in November 1999. On November 19, the -2F boosted the new, unmanned Shenzhou 1 spacecraft into low Earth orbit. Three more tests of the -2F followed, in which it launched Shenzhou 2 to 4 (with 2 carrying live animals). Then, on October 15, 2003, China joined the US and Russian space programs as the only ones with the capability to carry human beings (in this case, taikonaut Yang Liwei) into space.

Since then, the Long March-2F recorded a string of successes, sending Shenzhous 6 through 11 into orbit from October 2005 to October 2016.

ROBOTICS
Opportunity and Curiosity

NASA's bicentennial gift to the nation—the twin landings on Mars by Viking 1 and 2 in June and August 1976—represented decisive steps forward in planetary exploration. But surprisingly, the United States followed the success of the Vikings with a fallow period in Mars exploration, revived finally on July 4, 1997, when employees at the Jet Propulsion Laboratory (JPL) watched as another groundbreaking project reached the Red Planet. Launched on December 4, 1996, Mars Pathfinder combined a lander that touched down on Mars with a mobile buggy designed to drive itself over the Martian landscape. Actually a six-wheeled cart no bigger than a small suitcase, the 23-pound rover Sojourner (named for the prominent American abolitionist, Sojourner Truth) carried three cameras, as well as an alpha-particle x-ray spectrometer (APXS). Once free to roam the Martian flood zone where it touched down, Sojourner sent home 550 images and fifteen chemical analyses of rocks and soil. During its twelve-week journey, it traveled no more than 328 feet (100 meters) in all, never going farther than 39 feet (12 meters) from the lander.

The JPL team analyzed the data from the Pathfinder mission for several years, until in 2000 they proposed fabricating two bigger, more complex, and more capable rovers. Nature provided

Mars Exploration Rover (MER A/B – Spirit/Opportunity)
Mars Science Laboratory (MSL – Curiosity)

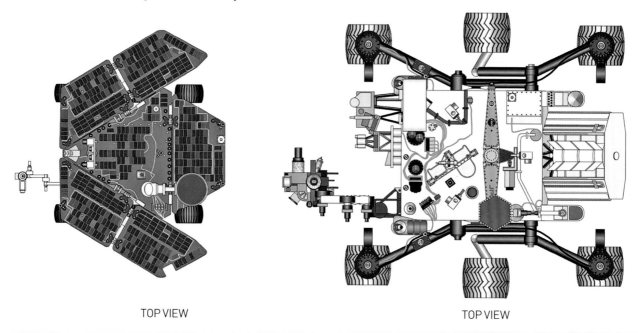

TOP VIEW   TOP VIEW

a powerful incentive and a clear timetable. The engineers and scientists seized on a planetary alignment predicted for August 2003 when the orbits of Earth and Mars brought them closer together than at any time in thousands of years. To speed up preparations, they borrowed heavily from the petite Sojourner: the same buggylike design with rocker-bogie suspension and six wheels, a cocoon of airbags to absorb the shock of landing, and a solar panel/battery power pack. The old and new designs diverged mostly in size. Sojourner measured only 7 inches (.18 meter) high and 26 inches (.66 meter) long; these next generation Mars Rovers stood 4 feet 11 inches (1.5 meters) tall, with a length of 5 feet, 2 inches (1.6 meters)—about the size of a golf cart. Each weighed 384 pounds (175 kilograms), or roughly seventeen times that of Sojourner.

About six years after Sojourner's mission ended, the new rovers left for Mars: one, launched on June 10, and a second on July 8, 2003, each aboard a Delta II rocket. Known formally as the Mars Exploration Rovers, they landed with a flourish. Encased in aeroshell capsules, they sped through the Martian atmosphere until just two minutes prior to impact. At that point, a parachute opened, airbags inflated, a retrorocket fired, and the rovers struck the surface and bounced repeatedly (twenty-eight times in one, twenty-six in the other) until coming to a stop. The first landed on January 4, and the second on January 25, 2004. NASA decided to name these twins by holding a national competition

among approximately ten thousand school children. The winners: Spirit and Opportunity.

JPL designed the missions of these two explorers to be distinct and separate. Spirit landed in Gusev Carter, a depression similar in size to Connecticut, that may have once been a lake. Opportunity set about its work on the other side of the planet, on Meridiani Planum, a plain about as big as Oklahoma with deposits of hematite—a mineral associated with liquid water. To evaluate what the rovers traveled over, the JPL team equipped the spacecraft with five instruments: a panoramic camera mounted on a five-foot mast, capable of high resolution, full-color, stereographic images; a microscopic imager to do closeups of soils and rocks; a miniature thermal emission spectrometer to scan the landscape and identify mineral deposits; a Moss Bauer spectrometer to measure the iron content in certain minerals (especially those linked with water in their formation); and an alpha-particle x-ray spectrometer to determine major elements present in the rocks and soils.

In the first phase of discovery, Spirit sent home images of a group of seven mounds on the landscape that the mission planners called the Columbia Hills (after the recently lost space shuttle *Columbia*). They named these topographic features after the lost Columbia astronauts: Anderson, Brown, Chawla, Clark, Husband, McCool, and Ramon. During its first seven months, Spirit went up to and over them, onto a plateau called Home

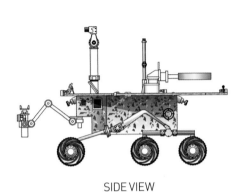

SIDE VIEW

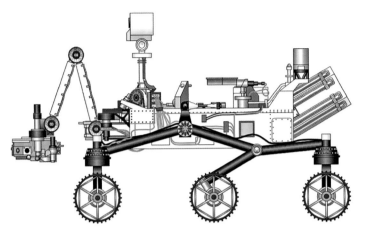

SIDE VIEW

Plate, over a ridge, and back to Home Plate. By 2007, it covered 4.5 miles, finding (among other things) that water did affect the mineral composition of some soils and rocks. Opportunity, meanwhile, sent home data about Martian craters, driving over the Eagle, Endurance, and massive Victoria crater and logging 6.5 miles through mid-2007. Its survey work confirmed that in the distant past, water flowed over and saturated parts of the planet, according to the texture evidence on rocks and minerals.

From the start, mission control at JPL struggled to keep the twin rovers mobile: the computer on Spirit rebooted itself repeatedly and at one point Opportunity was bogged down in soil for four weeks. Yet, built to last for three months, they went on for years. In the end, Spirit gave up first, becoming stuck in soft sand in 2009 and ending transmissions to Earth in 2010. Opportunity continues in operation, traversing the Endeavour Crater since 2011 and, within it, Marathon Valley to search out clay mineral deposits. By 2017, it moved on to Perseverance Valley (also in the Endeavour Crater). Meanwhile, JPL controllers learned to cope with the bitter Martian winters, the first of which inactivated Opportunity for four months. To minimize problems, they placed the spacecraft in sunny areas to keep it functioning during the dark months and they positioned it to benefit from the prevailing light breezes that blow the dust off its solar panels. Despite these and other vagaries, Opportunity still returns data more than 14 years after its deployment, having crossed more than 28 miles (45 kilometers) of terrain by 2017.

Ever since NASA began exploring Mars up-close—with the

Mariners in the mid-1960s, the Viking Landers in the mid-1970s, Pathfinder in the mid-1990s, and the Mars Exploration Rovers in the early twenty-first century—scientists have searched for signs of water. In fact, they consciously pursued their research guided by the principle, "Follow the Water," on the assumption that this indispensable ingredient preceded all life forms. And after these missions, they concluded that the geological record pointed inescapably to a former period of flowing water on the Red Planet. Their work led them to speculate about an ancient Mars (3.8 to 3.5 billion years ago) more like present-day Earth—much more wet and warm than the current Martian environment. Because the first microbes appeared on Earth in the same timeframe and under similar conditions, they wondered whether a parallel evolution might have happened on Mars. Translated into programmatic terms, these insights led the scientists to refocus their efforts from "Follow the Water" to a new campaign, "Seek Signs of Life."

This reassessment resulted in the Mars Science Laboratory and its Curiosity Rover. Its program managers hoped that Curiosity's technological advances over previous spacecraft increased the likelihood of finding evidence of organic compounds, the essential element in the basic soup of life.

Like so many other US planetary probes and robots—not least of which Spirit and Opportunity—JPL's engineers and scientists designed and fabricated the Mars Science Laboratory. Compared to the generations of Mars missions that preceded it, Curiosity represented a leap in size and complexity. If Sojourner could fit into a suitcase and the Mars Rovers looked like a golf carts, Curiosity's dimensions seemed suited to an automobile showroom: almost 10 feet (3 meters) long, 9 feet (2.7 meters) wide, 7 feet (2.2 meters) tall, and with a weight of roughly 1,982 pounds (899 kilograms). Additionally, it came equipped with a 7-foot (2.2-meter) long mast for picture-taking and for grasping rock and soil samples.

An Atlas V rocket launched Curiosity on November 26, 2011, and it landed in Gale Crater on August 5, 2012. It traveled to Mars embedded in an 8,583-pound (3.893-kilogram) vehicle that looked like a clamshell on a pie plate. Too heavy to imitate the landing method of the Mars Rovers, Curiosity began its mission with a new form of descent. Rather than falling to the surface in an aeroshell with limited control, Curiosity's protective spacecraft steered its course until it neared the landing zone. At that point, it activated its parachute, then its retrorockets, and lowered Curiosity on a tether (like that of a sky crane helicopter), finally dropping it on its target.

Although its basic features—a six-wheel drive, a rocker-boogie suspension, and cameras mounted on a mast—borrowed from the earlier rovers, its science payload really functioned as an on-site scientific workshop. Its sample analysis laboratory tested geologic materials and atmospheric elements; its x-ray diffraction instrument and alpha particle x-ray spectrometer differentiated minerals in the rocks and soil and determined their proportions; its mast held a hand lens imager that took extreme closeups, as well as a camera that took stills and video in high resolution, stereo, and color. Its ChemCam targeted laser light to vaporize thin layers of soil or rock and its radiation assessment detector measured the degree of radiation on the Martian surface. Its Mars descent imager captured high-definition video of the spacecraft's landing site, and the Dynamic Albedo of Neutrons instrument detected hydrogen (and potentially water) up to 3 feet (1 meter) below the surface.

Once inside of Gale Crater, Curiosity got to work. After about six years of exploration and 11 miles (17.7 kilometers) of travel, it reported significant findings. Curiosity found unmistakable signs that water once flowed on Mars, such as the discovery of an ancient dry stream bed about 4 feet deep and at least a few miles long. Its instruments further revealed that the planet once contained a thicker atmosphere with a greater abundance of water, suggested by the presence of isotopes of argon, hydrogen, and carbon. Additionally, the research uncovered methane in the atmosphere, the concentration of which actually increased ten times in just two months—tantalizing because living organisms expel methane (as do some chemical interactions between rocks and water).

Curiosity also found organic molecules in powdered rock samples—not in itself an indicator of present or past life on the planet, but a hint that the mixture of nutrients required for organisms may have existed at one time. And because the Mars Science Laboratory identified four of the mainstays of life—carbon, hydrogen, oxygen, and sulfur—scientists believed that the Red Planet may have once sustained microbial activity.

On the other hand, for those hoping to send humans to explore Mars, a red flag: the radiation levels found by Curiosity exceeded the amounts deemed safe for astronauts by NASA.

Chandra Telescope

Among NASA's Great Observatories (which also include Hubble, Compton, and Spitzer), the Chandra x-ray telescope took the longest path to completion. Its origins date to 1976, fully twenty-three years before its launch in July 1999. It began with a proposal to NASA from academic sources, followed in 1977 by a joint sponsorship offer from NASA's Marshall Space Flight Center and the Smithsonian Astrophysical Observatory at Harvard University. They named it Chandra after Nobel laureate Subrahmanyan Chandrasekhar (1910–1995), an Indian-born professor of astrophysics at the University of Chicago. His work concentrated on supernovas, neutron stars, and black holes.

As conceived of originally, Chandra represented a counterpart to Hubble: a spacecraft that benefited from occasional servicing missions from the space shuttles. Research on the instrument progressed during the 1980s, focusing on its mirrors and other technologies, after which fabrication began at Thompson Ramo Wooldridge (TRW), Chandra's prime contractor. But NASA Administrator Daniel Golden—who, under pressure from the Clinton administration to hold down costs, advocated "faster, better, cheaper" space systems—decided to reduce the scope of Chandra. The telescope's original twelve mirrors fell to eight, and its six scientific instruments declined to four. Moreover, instead of a low Earth orbit, mission planners decided to launch Chandra into a high elliptical one, about a third of the way (87,000 miles/140,000 kilometers) to the moon at its most distant point, where it had no possibility of visits from the shuttle, which flew only in low Earth orbit. Construction on the reimagined Chandra continued until March 1998 when TRW completed its overall assembly.

As the spacecraft came into being, a complex network of institutional relationships arose to support it. Marshall assumed overall project management under the supervision of the NASA Science Mission Directorate. But the contents of the missions became the responsibility of the Smithsonian Astrophysical

Chandra X-Ray Observatory (CXO)

FRONT VIEW

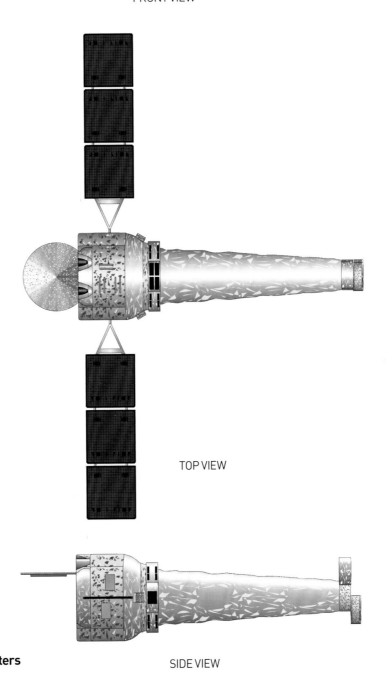

TOP VIEW

SIDE VIEW

0 1 2 3 4 5
meters

Spitzer Space Infrared Telescope Facility (SIRTF)

FRONT VIEW

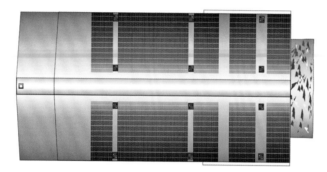

TOP VIEW

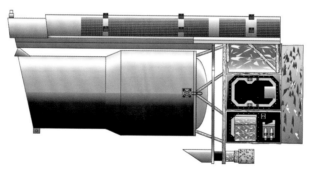

SIDE VIEW

meters

Observatory, in consultation with the world's science community. The Chandra X-Ray Center—a consortium of the Smithsonian, the Massachusetts Institute of Technology (MIT), and TRW—managed flight operations from its Cambridge, Massachusetts, location. And communications for Chandra relied on JPL's Deep Space Network.

Loaded into the payload bay of the shuttle *Columbia* for its launch on July 23, 1999 (during STS-93), Chandra represented the heaviest cargo ever carried by the space transportation system: 12,930 pounds (5,865 kilograms) itself, in addition to a 32,500-pound (14,740-kilogram) inertial upper stage (IUS)—a two-part solid fuel booster needed to push Chandra into the more distant loop decided on by NASA. With support equipment, the payload weighed 52,162 pounds (22,753 kilograms). And Chandra not only pressed the limits of mass, but of size as well. It filled the shuttle's cavernous hold; between the IUS and the 45-foot-long telescope, the mated stack measured 57 feet (17.4 meters), a mere 3 feet (0.91 meters) shorter than the shuttle's 60-foot (18.3-meter) cargo compartment.

Beneath its tapered cylindrical exterior, Chandra consisted of three distinct segments. The first held computers and communications systems that enabled contact between it and ground stations—in essence, a command and control center. It also carried a camera that oriented it, machinery that operated its solar panels, and propulsion rockets to steer it. The second portion—the instrument itself—differed from past x-ray observatories in the size and smoothness of its eight tubular mirrors. Bigger than any up to this point, the largest measured 3 feet (0.9 meter) long and 4 feet (1.2 meter) in diameter and their combined weight totaled roughly 2,000 pounds (907 kilograms). Additionally, to provide constant temperatures for more accurate observations, Chandra's designers integrated internal heating units in the reflective sheathing that covered the outside of the spacecraft. These innovations enabled viewing with eight times greater resolution and from twenty to fifty times more sensitivity than past x-ray telescopes. The third section carried science instruments that recorded the images witnessed by the telescope, in particular the high-resolution camera that captured pictures of high-energy occurrences, such as the death of stars.

Chandra's initial period of operations ended in 2004, but after extensions, it continues its mission more than eighteen years after its launch. It watches distant galaxies rocked by explosions, some of which contain vast black holes at their centers. Scientists also use Chandra to increase their knowledge of dark matter, an ill-defined substance that holds together the hot, x-ray emitting gases between the galaxies. Finally, Chandra observed many unprecedented events during its lifespan: the first sighting of

a binary black hole, the first sound waves detected from a black hole, a black hole in the Milky Way, and the most far-off x-ray cluster ever detected. As recently as August 2016, Chandra and its ESA counterpart, the x-ray XMM-Newton Observatory, discovered CL J1001 some 11.1 billion light years from Earth, consisting of eleven enormous galaxies evolving from a loose amalgamation into a coherent cluster—a transformation never seen before by astronomers. Producing about three thousand suns per year, CL J1001 gives scientists clues about the formation of these immense galaxy confederations, the biggest structures in the universe bound by gravity.

Spitzer Space Telescope

Once the first three of NASA's four Great Observatories—Hubble, Compton, and Chandra—went into orbit during the 1990s, only one remained: the Spitzer Space Telescope. Like the others, NASA planners gave Spitzer a unique mission based on the spectrum of light that it observed. The case for infrared, Spitzer's viewing specialty, took some time to build. As early as the 1960s, astronomers flew infrared telescopes mounted on huge balloons to the upper atmosphere. During the 1980s, Great Britain, the Netherlands, and the United States sponsored the Infrared Astronomical Satellite, the first of its kind. NASA attempted a shuttle-based infrared telescope in 1985, but the heat and small particles generated by the orbiter itself doomed the project. The breakthrough for infrared astronomy occurred in 1989 when the US National Research Council published a report giving it the highest priority for the 1990s.

NASA seized on this recommendation with gusto, proposing an instrument costing about $2.2 billion. However, the Clinton administration and NASA Administrator Daniel Golden, who made frugality the watchword of spaceflight, trimmed the original estimate to about $500 million. The Jet Propulsion Laboratory (JPL) received overall control of the project and the California Institute of Technology (Caltech) managed its science operations. Lockheed Martin and Ball Aerospace manufactured the spacecraft.

Relatively lightweight at 2,095 pounds (950 kilograms) due to its beryllium structure, when it flew into space on August 25, 2003, atop a Delta II rocket, it went by the straightforward name Space Infrared Telescope Facility. But by the end of the year, a public contest held by NASA resulted in the more pleasing Spitzer Space Telescope, which honored astrophysicist Lyman J. Spitzer (1914–1997) of Princeton University. Spitzer won the competition in part because of his longtime loyalty to the cause of flying observatories: he proposed a space-based telescope as early as the 1940s and championed Hubble during congressional hearings in the 1970s.

Because infrared astronomy involves the detection of heat radiation, Spitzer's main problem stemmed from the control of temperature emanating from the instrument itself, a potential source of contamination of sensitive readings. Engineers found an imaginative way to cool the telescope down to the required -450 degrees Fahrenheit (-268 Celsius). Rather than try to carry enough cryogen to do the entire job, its designers decided to immerse the instrument chamber in cryogen only at launch, after which, for the following five weeks, the entire system underwent exposure to the frigidity of space prior to starting the telescope's operation. This plan saved weight and cost and enabled a much longer mission by preserving the cryogen hauled on the flight.

But Spitzer needed to fend off infrared interference from another source—the Earth itself. To do so, it avoided a standard, circular orbit, and instead trailed behind our planet on its voyage around the sun. At that safe distance, its 33-inch (84-centimeter) telescope, coupled with three science instruments that housed large-format infrared detector arrays, operated effectively. During its long service life (initially 2.5, later increased to five, now stretched to more than fourteen years), Spitzer examined phenomena close by and far away. Nearer to Earth, it searched for telltale dust disks around neighboring stars (signs of possible planetary formation) and peered through curtains of cosmic dust (behind which new stars often took shape). Looking into the more distant past, the telescope spotted the black holes and galaxy collisions that breed ultra-luminous infrared galaxies, in addition to witnessing the birth of early and distant galaxies. From such observations, researchers discovered that even in the fury unleashed when supernovas exploded, polycyclic aromatic hydrocarbons—regarded as one of the fundamental ingredients of life—survived the cataclysm.

More recently, NASA teamed Spitzer with Hubble and Chandra and trained them on six of the biggest, most distant galaxy clusters ever seen, located about 13 billion light years away. Called Frontier Fields, the project ended its observations in 2016 and researchers began to analyze the data, which will enable astronomers to look as far back in space and time as present technology allows. During the following year, Spitzer made another discovery: the uneven glow of brown dwarfs (objects classified as smaller than the sun but larger than Jupiter) resulted from patchy clouds blown by powerful winds across the face of these distant bodies.

NASA will continue to fund Spitzer through early 2019. After that, the space agency expressed an interest in transferring control either to a university or to a private firm.

Juno

Eager to know more about Jupiter after the eight-year journey of the Galileo spacecraft, researchers at the Jet Propulsion Laboratory conceived of a second major mission to the giant of the planets. They named it after the Roman goddess Juno, who saw through a veil of clouds erected by her husband, the chief God Jupiter, to hide his infidelities. NASA wanted its Juno to do the same, although in a different context: to penetrate the haze and reveal the planet's true scientific nature.

Although they shared the same target, Juno and Galileo took significantly different programmatic and technical paths. JPL managed both projects, but instead of acting as the manufacturer itself (as in Galileo), in Juno it contracted with Lockheed Martin for design and fabrication. It also relied on an outside source to lead the scientific inquiry. Scott Bolton of the Southwest Research Institute, headquartered in San Antonio, Texas, acted as principal investigator, in collaboration with fifty-nine co-investigators. The objectives of the mission differed from Galileo's too. Juno's team concentrated on Jupiter itself rather than dividing their attention on the planet and its moons. And unlike Galileo that made thirty-five orbits in eight years, mission planners expected Juno to make thirty-seven revolutions of Jupiter in about one and a half years. They adopted a close-in, polar orbit lasting just fourteen days, which enabled scientists to scan the entire planet without risking damage to Juno's electronics from the radiation belt that hovered at Jupiter's equator. Flying at low altitude also allowed accurate measurements of its gravitational and magnetic fields, and it enabled Juno to pierce the cloud cover.

A squat, six-sided spacecraft from which three massive solar arrays radiated (the first ever used at such a long distance from the sun), the body of Juno stood 11.5 feet (3.5 meters) high and 11.5 feet in diameter. It weighed an imposing 7,992 pounds (3,625 kilograms) at launch. The solar arrays measured 29.5 feet (9 meters) long by 8.7 feet (2.65 meters) wide—roughly 66 feet from tip to tip, longer than a travel coach bus and the biggest of any deep space probe to date.

Juno's path to Jupiter lasted a little longer than five years. It lifted off from Cape Canaveral, Florida, on August 5, 2011, aboard an Atlas V (with five solid rocket boosters and a Centaur upper stage). After conducting deep space maneuvers in August and September 2012, Juno flew by Earth for a gravity-assist in October 2013 and arrived at the giant planet on July 4, 2016. NASA anticipated an end to the mission after nineteen months, when, in February 2018, Juno will de-orbit into Jupiter.

High expectations accompanied Juno. Scientists believed that understanding the origins and evolution of the mighty

Juno Orbiter Probe

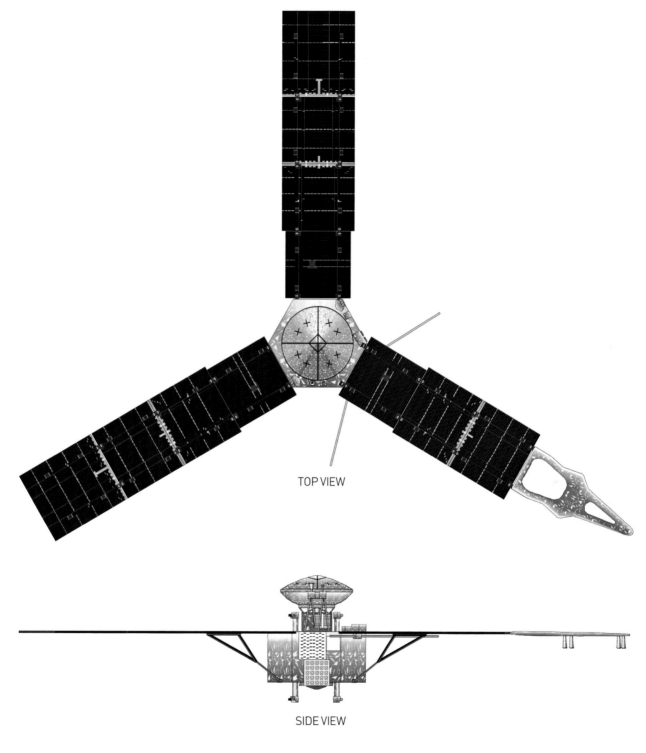

TOP VIEW

SIDE VIEW

Deep Impact Probe

TOP VIEW

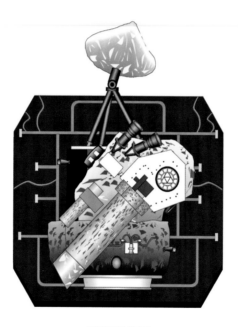

FRONT VIEW

meters

planet might serve as a Rosetta stone for the processes that contributed to the formation of the solar system itself. Additionally, they hoped to investigate its planetary core, map its magnetic field, gauge the extent of water in its interior atmosphere, and assess its auroras.

Its suite of instruments collected data about these and other Jovian mysteries with each sweep over the planet. Technicians placed on board gravity science instruments and magnetometers to learn about Jupiter's inner structure and its gravity and magnetic fields; a microwave magnetometer to assess its interior atmosphere and determine how much water (and oxygen) lay inside; equipment to sample the particles, plasma waves, and electric fields in an effort to understand the relationship between the magnetic field, the atmosphere, and the auroras; infrared and ultraviolet cameras pointed into its atmosphere and its auroras; and another camera (nicknamed JunoCam) for closeup color images.

NASA administered Juno—like New Horizons before it—under the New Frontiers program, devoted to medium-sized missions costing up to $1 billion.

Deep Impact

Only so many flight options present themselves to researchers planning space science missions. One classic method sends probes to fly past a target; a second one to orbit it; a third (more complex and subject to failure) for soft touch downs; and a fourth to drive over the terrain.

The fifth technique—intentional crash landings—occurred often during Soviet and US robotic missions in the late 1950s and 1960s. It recurred early in the twenty-first century, when an international movement of scientists made a concerted effort to expand their understanding of the origins of the solar system. The European Space Agency (ESA) mounted a mission to fulfill this goal called Rosetta, a cometary probe launched in 2004. Rosetta did not collide with anything, but instead gathered data with an orbiter and a soft lander.

The United States' answer to Rosetta, called Deep Impact, harkened back to the Ranger moon probes of an early space age—all of which ended their flights by smashing into the lunar surface. But the Deep Impact team proposed a hard landing like no other. Engineers and scientists at JPL—in cooperation with their counterparts at the University of Maryland—conceived of a bold mission in which controllers aimed their spacecraft at the nearby Tempel 1 comet, putting the two on a collision course. They hoped that Tempel 1's mass—45 square miles (117 square kilometers) in size—enabled it to absorb the blow without disintegrating. They also hoped that when Deep Impact struck,

it generated enough force to penetrate the comet's surface and reveal data about its origins—and in so doing, the origins of the planets as well.

The JPL-Maryland partners conceived of the mission between 1999 and 2001, after which they contracted with Ball Aerospace to fabricate and instrument the spacecraft. Actually, Ball built two separate but conjoined machines in one: a bigger flyby vehicle and a smaller impactor. The main spacecraft was about the size of an average sports utility vehicle with a mass of 1,430 pounds (650 kilograms). It maneuvered to its target by a group of hydrazine-fueled thrusters. It carried just two bundles of equipment to record the massive crash: a high-resolution Instrument—an 11.75-inch (30-centimeter)-diameter telescope, an infrared spectrometer, and a multispectral camera—focused tightly on the point of contact; and a medium resolution instrument—a 4.75-inch (12-centimeter)-diameter telescope—for a wide-angle perspective on the materials ejected outward from the collision site. The impactor itself measured just 39 inches by 39 inches (99 centimeters by 99 centimeters), but weighed a surprising 850 pounds (372 kilograms). Of that, 249 pounds (113 kilograms) consisted of a space-based battering ram consisting of copper plates machined into a semi-spherical shield. It carried a telescope similar to the smaller one on the flyby spacecraft. Augmenting these on-site instruments, the mighty Spitzer Space Telescope overlooked the Tempel 1 blast from a distance.

Deep Impact left Cape Canaveral, Florida, on a Delta II rocket on January 12, 2005, and after traveling about 83 million miles (134 kilometers) over five months, it came close enough to Tempel 1 to begin taking a stream of photographs. In July, mission controllers reoriented the flyby spacecraft for the impactor's descent, which they initiated on July 3. The next day, propelled by battery power, the impactor made three targeting maneuvers before it dove headlong into Tempel 1 at 23,000 miles per hour (37,000 kilometers). The July 4 pyrotechnics produced an explosion equivalent to the force of 9,600 pounds (4,350 kilograms) of trinitrotoluene (TNT). It created a spectacle worthy of the holiday, digging an immense crater and raising a gigantic plume visible from the Earth (although faintly) with the unaided eye. As the cloud of particles rose, Deep Impact's bigger flyby telescope, as well as Spitzer, peered into the massive pit and found some expected materials, such as ice and silicates. But they also discovered the unexpected: hydrocarbons, carbonates, iron-bearing compounds, and clay. And despite the immense flash at ground zero, further research hinted that Deep Impact did not enter the comet with the force anticipated; the impactor went in a few times its own depth only because it encountered highly porous material—not the solid crust predicted.

After the big explosion, NASA headquarters gave the JPL scientists approval to extend the mission, and the flyby spacecraft went on to observe several other comets: Hartley 2 in 2010, Garradd in 2012, and ISON in 2013. In the end, onboard computer malfunctions compelled NASA to abandon Deep Impact in September 2013.

New Horizons

At the beginning of the twenty-first century, momentum gathered for a mission to Pluto, the little-understood planet at the distant frontier of the solar system. In 2005, the mystery of Pluto deepened when the Hubble Space Telescope discovered four small previously unknown moons circling it: Nix, Hydra, Styx, and Kerberos. The National Academy of Sciences added to the allure when it gave Pluto and the Kuiper Belt (a region of thousands of celestial objects orbiting the sun beyond Neptune) the highest priority for exploration.

But even without these recent attractions, Pluto represented a well-known list of exciting oddities. Discovered by American astronomer Clyde Tombaugh in 1930, it represented the only binary ("double") planet in the solar system; the only planet moving in space through a sea of other binary bodies; the only planet made neither of gas (the common ingredient of the four big outer planets) nor rock (the most abundant part of the four small inner planets), but of ice. It also embodied a planet with an eccentric orbit; an escaping atmosphere; an enormous moon, Charon; and a planet no longer considered such by the International Astronomical Union, but rather a "dwarf planet". It is just 1,400 miles (2,253 kilometers) in diameter, compared to Earth's 8,000 miles (12,874 kilometers).

New Horizons—about the size of a grand piano and comparatively light at 1,054 pounds (478 kilograms)—went into orbit atop an Atlas V launcher on January 19, 2006. It flew unfettered by solar arrays (due to its distance from the far-off sun) and instead received power from a nuclear radioisotope thermoelectric generator (RTG) that converted heat from its plutonium-238 pellets into electricity. Planetary mechanics prescribed New Horizons' launch window: during a three-week period at the start of 2006, the spacecraft took the opportunity of a favorable alignment to fly straight to Jupiter. Propelled by the gravity-assist of the mighty planet, it accelerated by 9,000 miles per hour (14,482 kilometers per hour) and cruised at 41,000 miles per hour (65,997 kilometers)—the fastest speed achieved by any spacecraft to that time. This slingshot effect carried New Horizons to its target three years sooner than by a direct route from Earth.

New Horizons made its closest contact with Jupiter on February 28, 2007. Researchers received seven hundred data sets about the Jovian encounter in June 2007, including observations about its atmosphere, rings, and closest moons (Io, Ganymede, Europa, and Callisto).

Then came the main event of the mission. On July 14, 2015, Pluto became the last planet to receive an initial visit from a spacecraft, when New Horizons reached the nearest point on its flyby of the distant enigma. The voyage overturned many of the common assumptions about Pluto; its complexity (and that of its moons) far exceeded prevailing expectations. In fact, Pluto held great surprises. Because of New Horizons, researchers believed that beneath the planet's surface an ocean of water or ice exists, and they thought that the same might be true of Charon, although in the distant past. Additionally, some portions of the planet's exterior probably came into being far more recently than anticipated. Scientists also discovered that at least some of Pluto's moons appear to be about the same age, leading them to conclude that they originated long ago, after a collision between the planet and another object in the Kuiper Belt. Finally, the expected atmospheric escape rate appeared to be lower than predicted.

After passing Pluto, NASA sent New Horizons on an extended mission through the Kuiper Belt, targeting an object known as 2014 MU69 for an encounter in 2019. With this voyage, New Horizons will complete mankind's preliminary survey of the solar system.

James Webb Space Telescope

NASA made an unlikely choice when it named the world's biggest space telescope after James E. Webb, just as President John F. Kennedy made an unlikely choice in naming him as NASA administrator, a position he held from February 1961 to October 1968. A lawyer, Webb had no training as an engineer or scientist. But he knew the ropes of government, having served as a congressional aide, under secretary of the Treasury Department, under secretary of State Department, and director of the Bureau of the Budget. Four months into Webb's job at NASA, the president announced the mission to the moon, and this decision gave Webb the assignment that made him famous. Whether in the halls of Congress or in other Washington, DC, corridors of power, he became the chief protector and protagonist of the Apollo program, and this ultimately earned him the association with the great telescope that will bear his name.

This new instrument—often referred to as the successor of the Hubble telescope—took many years to develop, as did Hubble itself. Initial discussions for what at first became known as the Next Generation Space Telescope (NGST) occurred as early as 1989, but in 1993—the same year that astronauts corrected

New Horizons Probe

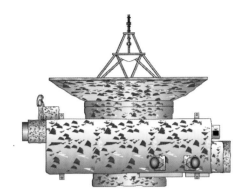

FRONT VIEW

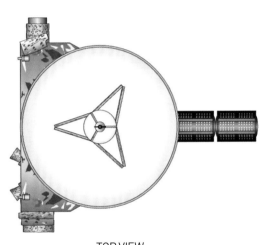

TOP VIEW

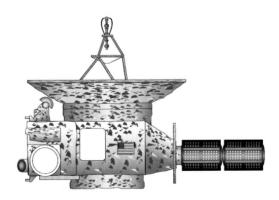

SIDE VIEW

0   1   2   meters

James Webb Space Telescope (JWST)

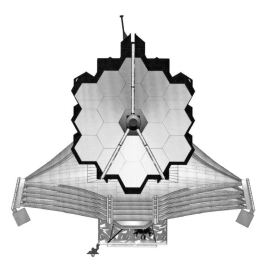

FRONT VIEW

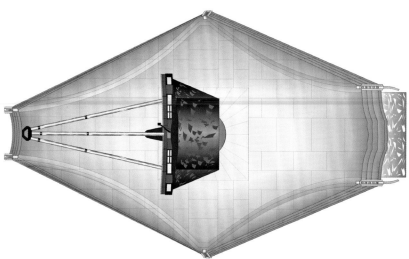

TOP VIEW

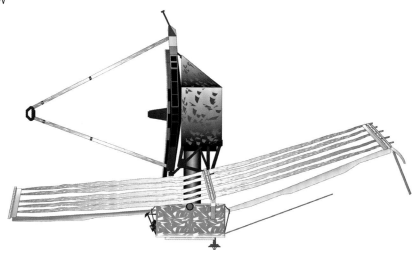

SIDE VIEW

0 1 2 3 4 5
meters

the optics of Hubble in a series of hair-raising spacewalks—the astronomical community began to define the upcoming telescope and its objectives. They decided that NGST ought to concentrate on a broad spectrum, from long wavelength visible light through mid-infrared. By 1996, NASA Goddard Space Flight Center and three corporations prepared feasibility studies.

The project accelerated from 1997 to 2001, during which time astronomers further described the specific characteristics desired of the NGST. In the meantime, the European and Canadian Space Agencies (ESA and CSA) joined the project. In 2001, Thompson-Ramo-Wooldridge (TRW)/Ball Aerospace competed with Lockheed Martin to become the prime contractor. In the end, NASA headquarters put Goddard in charge of the NGST and issued the project's birth certificate in September of 2002 when it awarded the mirror and spacecraft work to TRW. At the same time, the space agency announced a projected launch date of 2010 and named the instrument the James Webb Space Telescope.

With that, the full weight of this ambitious project descended on NASA, Goddard, TRW, its subcontractors, and the Space Telescope Science Institute of Baltimore, Maryland (that managed Webb's flight operations and science proposals, as it did for Hubble). A four-year period of detailed design work started in 2003, during which time ESA offered its powerful Ariane 5 rocket for the liftoff. Some unwelcome news came to light in 2005: the telescope proved to be too heavy, which the engineering team resolved by switching from a cryostatic (active) cooling system that governed onboard temperatures to a cryocooler (passive) approach that chilled by exposure to space. But the management side of the program proved to be less amenable to correction. Revised figures from Northrop Grumman, which acquired TRW in 2002, reported development costs rising from under $2 billion to $3.5 billion, and launch dates falling back, first to June 2013 and later to June 2014.

Congress did not take the news lightly, even less so in 2011 when the House and Senate met to decide Webb's fate. NASA presented a disturbing new baseline, now requesting an $8.7 billion budget and a launch in May 2020. The House of Representatives Appropriations subcommittee voted to terminate the project outright. But the vigorous arguments of Senator Barbara Mikulski of Maryland (home state of NASA Goddard and the Space Telescope Science Institute) persuaded the US Senate to approve the new targets proposed by the space agency. By early 2014, Webb regained its momentum; 97 percent of the telescope's mass and all four of its main science instruments arrived at Goddard for testing and integration. Construction on Webb ended in 2016.

In fairness to NASA and its collaborators who bore the brunt of the congressional scrutiny, Webb surpassed almost

every other spacecraft in its scientific potential, and its raw size and anticipated image clarity underscored its claim to preeminence. Its largest feature consisted of a multilayered sunshield almost as big as a regulation tennis court, measuring more than 72 feet (22 meters) in length by nearly 33 feet (10 meters) in width. It looked like a long, octagonal screen made of lightweight reflective material, designed to protect the telescope from the light and heat generated by the sun and Earth. (The cooling of the spacecraft assumed great importance because infrared telescopes required extreme cold for accurate readings). Additionally, the orbit of Webb around the sun at 1 million miles (1.5 million kilometers) from Earth enabled the sunshield to stay aligned with the Earth and the sun and to block out much of their radiation, therefore keeping temperatures down.

Towering vertically over this protective surface stood the primary mirror, a structure made of beryllium, coated in gold, and composed of eighteen smaller hexagonal mirrors that together spanned 21 feet (6.5 meters). The mirror's focal length measured 431 feet (131.4 meters). Webb carried four main instruments: a near infrared/visible range camera, a mid-infrared camera, a multiobject spectrograph, and a wide field spectrograph.

Scientists found it hard to contain their excitement as they waited for the five- to ten-year mission of the Webb telescope to unfold. They hope to discover evidence not just of the evolution of the universe, but of its origins. The most distant galaxies and stars—formed up to 13.5 billion years ago at the Big Bang—now seem within the grasp of astronomers. With the combined tools of infrared and Webb's immense mirror, scientists expect to penetrate the vast dust clouds that until now shrouded the birth pangs of stars and planets, even obscuring them to the Hubble telescope. And Webb also holds out the prospect of finding other planets that, like Earth, bear the constituents of life.

BepiColombo

Although the European Space Agency (ESA) member states normally operated among themselves in a cooperative and collaborative spirit, at times competition with other space agencies acted as an extra motivator. For example, NASA became the first space power to target Mercury for exploration, sending Mariner 10 there on flybys in 1974 and 1975, followed by MESSENGER to do three more of the same in 2008 to 2009, and then to orbit the planet from 2011 until it crashed landed in April 2015.

In considering these achievements, the Europeans may have felt a pang of rivalry when they remembered that the calculations of a prominent Italian mathematician—Professor Giuseppe ("Bepi") Colombo (1920–1984)—inspired Mariner 10's complex orbital path.

Accordingly, after initial reviews starting in 1993, ESA's governing council considered a Mercury mission known as BepiColombo and approved it in October 2000. ESA also showed its serious intentions in another way. While NASA produced MESSENGER as part of its low-cost Discovery program, eventually spending about $280 million, ESA, in contrast, categorized its trip to Mercury as a top priority cornerstone project, estimated to cost (on completion) more than $1.2 billion dollars.

Like MESSENGER, the European entrant in the Mercury sweepstakes will pursue a time-consuming and circuitous route to the planet nearest the sun. Indeed, like NASA's visitors before it, BepiColombo will find the planet a challenge to observe and hazardous to approach due to its scorching temperatures—at times in excess of 662 degrees Fahrenheit (350 Celsius). The Europeans planned to get there with nine gravity-assists: one from Earth, two from Venus, and six from Mercury itself. After its anticipated launch aboard an Ariane 5 ECA in October 2018, ESA authorities expected the probe to take seven years to arrive on site, following which it will pursue its mission until May 2027; or if extended, until May 2028.

BepiColombo actually consisted of three spacecraft in one: the Mercury Planetary Orbiter (MPO) fabricated for ESA by Astrium, a subsidiary of EADS (European Aeronautic Defence and Space Company); the Mercury Magnetospheric Orbiter (MMO), made by ESA's partner in the project, the Japanese Aerospace Exploration Agency (JAXA); and the Mercury Transfer Module (MTM), an ESA vehicle designed to convey MPO and MMO to their destination. The MTM served as the base of the tripartite structure, providing propulsion for the cruise phase and braking for its approach to Mercury. The MPO, shaped like a rectangular box and weighing 2,513 pounds (1,140 kilograms), measured almost 8 feet (2.4 meters) by 7.2 feet (2.2 meters) by 5.5 feet (1.7 meters), not including its solar arrays. Its suite of eleven instruments—seven spectrometers, an accelerometer, an altimeter, a magnetometer, and a radio science experimental unit—will enable a sweeping panorama of the planet's composition. The MMO, a squat octagonal cylinder, measures 6.2 feet (1.9 meters) by 3.6 feet (1.1 meters) and weighs 635 pounds (288 kilograms). Its five instruments will analyze the plasma and neutral particles emanating from the planet itself, its magnetosphere, and the interplanetary solar wind; they will also assess magnetic fields, plasma waves, the atmosphere, and dust.

Once at Mercury, the operational plan called for the MTM to be jettisoned and the still conjoined MPO and MMO to enter a polar orbit. The MPO separated itself at this point and flew to a lower altitude by means of chemical propulsion.

Many aspects of Mercury continue to puzzle scientists. BepiColombo's researchers hoped to resolve some of them,

such as the reasons why Mercury's density far exceeded that of other rocky planets and the moon. They also hoped to discover the nature of the planet's core (solid or liquid?); whether tectonic activity occurs on it; why—unlike Venus, Mars, and the moon—a magnetic field surrounds it; and why spectroscopes fail to detect iron there even though scientists believe that it constitutes a major element on the planet.

Provided that BepiColombo achieves its objectives, it will further distinguish ESA as one of the world's most adroit practitioners of space science.

Herschel and Planck

For those who doubt the practicality of international cooperation in an age of intensifying nationalism, the European Space Agency's (ESA's) Herschel and Planck Space Observatories offer a case study of success. In this instance, twenty-two sovereign countries shared their resources to design, fabricate, and operate two of the most advanced space projects ever conceived.

Differing in their missions, these two spacecraft also bear the names of two very different men. Sir William Herschel (1738–1822) grew up in Hanover, Germany; moved to England in 1757; and around 1766 became interested in optics. He taught himself to grind lenses and to make some of the finest telescopes of the age, with which he surveyed and cataloged the night sky and eventually discovered the gas giant Uranus. He concentrated on finding nebulas, cataloging 2,500 of them. If Herschel embodied experimentalism, Max Planck (1858–1947) represented the archetypal theorist. The son of a distinguished German legal scholar, Planck lived his first years in Kiel and then moved with his family to Munich. He earned his doctorate at the University of Munich when he was twenty-one and taught mainly at the University of Berlin, where over many years he evolved (with others) the revolutionary quantum theory of physics.

Herschel and Planck concerned themselves with origins. Herschel's planners devised the most powerful infrared telescope ever conceived, one that covered all wavelengths—from far infrared to submillimeter. With it, they hoped to detect water in distant parts of the universe and to observe the birth of galaxies, stars, and dust clouds. In contrast to Herschel, Planck's scientists trained a set of exceptionally sensitive instruments on what remained of the radiation emitted just after the creation of the universe, enabling them to identify constituents that formed soon after the Big Bang, such as dark matter and dark energy. Planck's forerunners included two NASA missions: the Cosmic Background Explorer (COBE), launched in 1989, and the Wilkinson Microwave Anisotropy Probe, sent aloft in 2001.

BepiColombo (interplanetary cruise configuration)

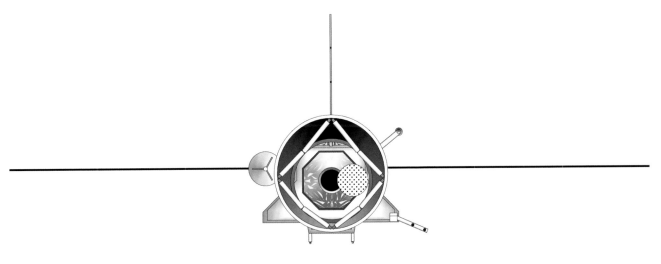

TOP VIEW

SIDE VIEW

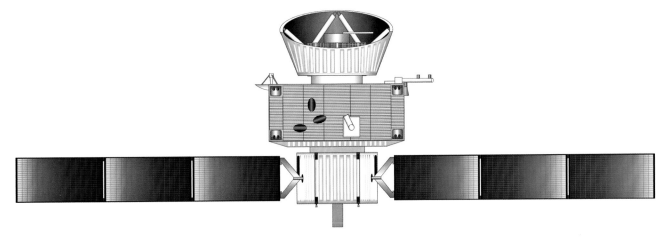

FRONT VIEW

Herschel Space Observatory

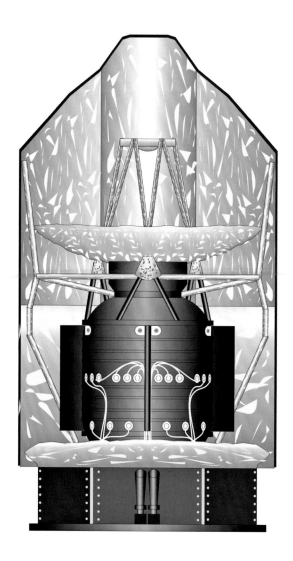

FRONT VIEW

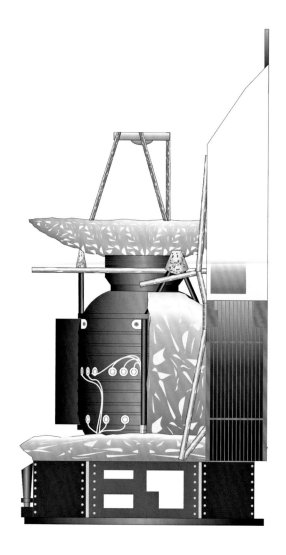

SIDE VIEW

0   1   2
meters

Planck Surveyor

FRONT VIEW

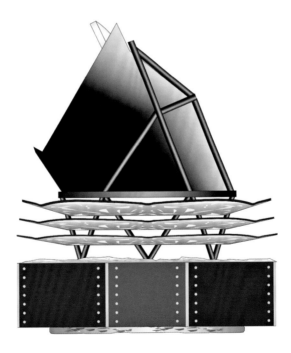

SIDE VIEW

0   1   2
meters

According to ESA's Director General Jean-Jacques Dordain, the Herschel and Planck spacecraft represented "the most complex science satellites ever built in Europe." Herschel looked like a traditional, canister-shaped telescope. Planck, on the other hand, resembled an eight-sided platform topped by a tall, pompadour-style haircut (the primary mirror). Herschel totaled 7,496 pounds (3,400 kilograms) at launch and measured 25 feet (7.5 meters) long and 13 feet (4 meters) in diameter. Its telescope weighed 694 pounds (315 kilograms) and it became the biggest optical instrument deployed in space to its time, with a 132-inch (3.35-meter) main mirror—about half again larger than Hubble's. Planck weighed 4,300 pounds (1,950 kilograms) at launch and carried a payload—including a 6-foot-by-5-foot (1.9-meter-by-1.5 meter) primary mirror—of 452 pounds (205 kilograms). Overall, it measured 13.8 feet by 13.8 feet (4.2 meters by 4.2 meters). Fabrication of Herschel and Planck occurred in the factories of the Franco-Italian satellite manufacturer Thales Alenia Space.

Herschel carried three main pieces of scientific equipment: the heterodyne instrument (a spectrometer) for the far infrared (HIFI); the photodetector array camera and spectrometer; and the spectral and photometric imaging receiver. Planck's telescope gathered light from the cosmic microwave background and focused it on two receiving points: the high-frequency and the low-frequency instruments. Detectors on these machines enabled Planck to transform the microwave and radio light that it collected into sweeping maps of the night sky.

An Ariane 5 ECA rocket lifted Herschel and Planck into space on May 14, 2009. About twenty-six minutes after the launch, ESA mission control in Darmstadt, Germany, released Herschel, and about two minutes later, Planck. About two months later, both entered the Lissajous orbit around the second Lagrangian point of the sun-Earth system (located about 933 million miles, or 1.5 million kilometers from Earth) in the direction opposite to the sun.

Herschel's journey ended in June 2013, four years after it began. Its supply of cryogenic helium—necessary to keep the spacecraft's instruments at a temperature near absolute zero—ran out, and controllers ended its mission. Its designers had expected a lifespan of three and one-half years. Among its many discoveries, Herschel found enormous quantities of water vapor in planet-forming disks of gas and dust surrounding newborn stars. Scientists believed that the water contained in these disks might constitute the seedbed of planetary oceans, like those on Earth. Herschel also observed dust and gas filaments in the Milky Way, structures capable of forming into the solid cores of new stars. Herschel made thirty-five thousand scientific observations and captured twenty-five thousand hours of science

data during its lifespan—an archive organized and preserved for future scientific inquiries.

Controllers at ESA terminated Planck's voyage in October 2013 when its liquid helium coolant diminished to the point where it compromised the spacecraft's effectiveness. But before that time, it completed five full sky surveys—three more than originally planned. The resulting data gave the clearest portrait yet of the young universe (about 380,000 years after the Big Bang) and offered revised proportions of its key ingredients. Because of Planck, the estimates now stood at 68.3 percent dark energy, 26.8 percent dark matter, and 4.9 percent ordinary matter (compared to 72.8 percent, 22.7 percent, and 4.5 percent prior to Planck).

Because of their exceptional sophistication, complexity, and success, Herschel and Planck represented the culmination of ESA's science program, recognized in 2015 when the American Institute of Aeronautics and Astronautics (AIAA) awarded the Herschel and Planck teams the prestigious AIAA Space Systems Award.

LISA Pathfinder

The European Space Agency (ESA) earned an enviable reputation on flagship spacecraft such as Herschel, Planck, and BepiColombo. It also sponsored smaller but no less impressive projects. One of these, first proposed in 1998, involved the measurement of gravitational waves. It bore the acronym ELITE (the European Laser Interferometer Space Antenna Technology Experiment). An expanded version of ELITE appeared in 2000 before ESA's Science Programme Committee, which added a separate vehicle—the Darwin Pathfinder—to the ELITE mission. Scientists and engineers planned to use ELITE to launch test objects sensitive to gravitational waves into space, flying them in formation with Darwin as a demonstrator. The committee gave the go-ahead and it became the second entrant in ESA's new Small Missions in Advanced Research Technology (SMART) series. ESA authorities ultimately decided to eliminate Darwin (at least in name) and rechristened the project LISA Pathfinder.

Although the immediate goal of the LISA Pathfinder team became the isolation and detection of gravitational waves, they also aimed for a more historic objective. Albert Einstein predicted the existence of these minute ripples in the fabric of space, caused by immense cosmic forces such as supernovas or the merger of supermassive black holes, and LISA Pathfinder went into orbit one hundred years and one day after the publication of his general theory of relativity, in which he posited this phenomenon. LISA Pathfinder's researchers wanted to not only verify Einstein's theory, but in so doing to deepen the meaning of general relativity and to more accurately comprehend the influence of cataclysmic events

LISA-Pathfinder Probe

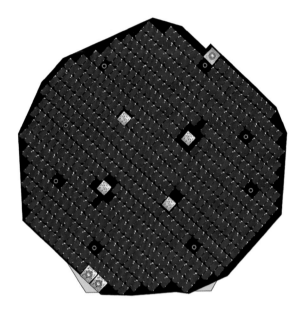

TOP VIEW

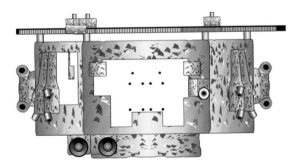

SIDE VIEW

meters

Rosetta Spacecraft
With Philae Lander integrated

TOP VIEW

FRONT VIEW

on the overall tapestry of the universe. Unfortunately for ESA, in 2015 a group of American scientists using the ground-based Laser Interferometer Gravitational-Wave Observatory (LIGO) confirmed Einstein's hypothesis by witnessing gravitational waves for the first time. But LISA Pathfinder held a big advantage; flying in space, rather than confined to Earth, enabled it to "see" the gravitational effects of these violent convulsions among the cosmos more effectively than LIGO.

Shaped like an octagonal pillbox, LISA Pathfinder measured only 2.75 feet (0.85 meters) tall and 6.9 feet (2.1 meters) in diameter. Heavy for its size, the spacecraft weighed 4,200 pounds (1,910 kilograms) at launch—of which just 276 pounds (125 kilograms) constituted the payload. In fact, unlike most other spacecraft, no real distinction existed between LISA Pathfinder's cargo and the vehicle itself. They behaved as a single unit; for instance, the science instruments maintained attitude control.

LISA Pathfinder carried two main components: the LISA technology package and the NASA-made disturbance reduction

system. The greatest challenge to LISA Pathfinder consisted of separating the measurement of gravitational forces from extraneous influences. To do so, the LISA technology package suspended one gold and one platinum cube—weighing about 4.4 pounds (2 kilograms) each—nearly 15 inches (38 centimeters) apart, enabling them to move freely in relationship to one another as they responded to the forces of gravity. As the spacecraft flew, the optical interferometer on board the spacecraft tracked their relative acceleration and detected the distance between them down to a trillionth of a meter. Meanwhile, the disturbance reduction system protected the purity of the data by employing a micro-propulsion system with two clusters of colloidal thrusters, as well as a control software to curb the effects of drag.

LISA Pathfinder entered space on December 3, 2015, aboard one of ESA's recently developed Vega rockets, a small launcher designed to carry multiple missions. At first, the satellite assumed an elliptical pathway around the Earth of roughly 124 miles (200 kilometers) at its closest point and nearly 957 miles (1,540 kilometers) at its most distant. Then, firing its own propulsion system, it raised its orbit six times, so that six weeks after liftoff it arrived at the sun-Earth Lagrangian point, nearly 1,000,000 million miles (about 1,609,300 kilometers) from where it started. LISA Pathfinder began operations on March 1, 2016.

During its scheduled one-year period of activity, it exceeded ESA's expectations. The LISA Technology Package recorded the fluctuating distance between the cubes one hundred times more accurately than expected. The technologies demonstrated by this satellite gave ESA scientists confidence to plan for a projected 2034 launch of three such satellites, positioned more than 621,000 miles (1 million kilometers) apart—in effect, a functioning, space-based gravitational observatory. ESA deactivated LISA Pathfinder on June 30, 2017.

Rosetta

Like its American cousin Deep Impact, the European Space Agency's (ESA's) Rosetta spacecraft also chased comets. But they differed in two main respects: ESA aimed for a far more distant point (the Comet Churyumov-Gerasimenko) and Rosetta attempted a combined orbital mission and landing, not a collision like Deep Impact. And, although launched within ten months of each other (Rosetta first on March 2, 2004), a gap of over nine years separated the crash of Deep Impact from the succeeding work of Rosetta.

Also like Deep Impact's team, Rosetta's creators (some of whom first contemplated the mission in the 1970s) pursued their research with the belief that comets predated the planets and chose a name that reflected the objectives of their project. Rosetta refers to a slab of rock discovered in Rashid (Rosetta) Egypt in 1799, on which ancient carvers inscribed the same text in hieroglyphics, early Egyptian, and Greek script. Just as the multilingual tablet enabled scholars to break the mysterious written code of the pharaohs, modern researchers hoped to unlock secrets about the origins of the solar system with Rosetta.

Rosetta traveled slowly to its destination. Its Ariane 5 launch vehicle lacked the power for a direct flight to Churyumov-Gerasimenko because of the distance to be covered—roughly 291 million miles (469 million kilometers)—and due to the mass of the spacecraft, about 6,600 pounds (3,000 kilograms). To compensate, it sent Rosetta on successive gravity-assist flybys of the Earth in 2005, 2007, and 2009, adding one past Mars in 2007. Rosetta followed a highly elliptical orbit that brought it closest to the sun as it neared the Earth and Mars, and farthest out when it flew beyond Jupiter. Then, between June 2011 and January 2014, as it flew farthest from the Earth (where solar power waned) ESA mission control placed the spacecraft into hibernation, in which

all functions but the onboard computer and the thermal control system went off. As the probe returned to the inner solar system, the sun's power became more abundant and Rosetta reawakened after 957 days in suspended animation.

The spacecraft that ESA resuscitated consisted of two distinct parts. On the exterior of the orbiter—a big aluminum box 9-by-7-by-7 feet (2.7-by-2.1-by 2.1 meters)—technicians mounted a large communications dish on one side, and on the other a squat, six-sided lander (named Philae) weighing only 221 pounds (100 kilograms) and measuring just 3.3-by-3.3-by-2.6 feet (1-by-1-by-.8 meters). Rosetta bristled with instruments: eleven on the orbiter and nine on the lander, including four different spectrometers, a grain impact analyzer, and (on both spacecraft) comet nucleus sounding systems and imaging systems.

Once revived, Rosetta still had a 5.6-million-mile (9-million-kilometer) voyage before reaching the comet. Between May and August 2014, a series of braking maneuvers brought it into close enough range so that Churyumov-Gerasimenko's odd, dual-lobed shape became evident. The spacecraft arrived there on August 6, 2014, and began its orbital pattern, flying between 6.2 and 18.6 miles (10 and 30 kilometers) from the surface. Then on November 12, Rosetta released Philae, which fell to the comet's surface gradually, without power or guidance. When it touched down, harpoonlike probes jabbed into the comet's surface, but failed to grab hold, causing Philae to bounce and land three times before coming to rest on an unexpected part of the comet. There it pursued its science mission for the planned 2.5 days. Philae sent home sweeping views of the comet's landscape, high-resolution images of its surface, an analysis of its exterior composition and gases, and data on the structure of its interior (discovered when Philae sent radio signals through the nucleus to the orbiting Rosetta). With that, Philae's battery ran down because of a lack of sunlight to recharge it, and the lander ceased operations. It came back to life partially when lighting conditions improved.

Rosetta's orbiter continued to circle Churyumov-Gerasimenko after Philae's landing, collecting data for nearly two years. It observed terrains intense with activity, giving scientists the opportunity to witness fast-evolving cometary geology unfold before their eyes. Finally, on September 30, 2016, mission control sent signals that put Rosetta on a collision course with Churyumov-Gerasimenko; rather than wait for its power to dwindle, the project's leaders decided that a final, instrumented descent offered a chance to measure the comet's gas, dust, and plasma environments right near the surface and to take some final, high-resolution images just before impact.

SELECTED SOURCES AND FURTHER READING

The following represent a sampling of the sources consulted for this book.

I. The Internet

A cluster of authoritative and trustworthy websites constituted some of the main resources for this book:

1. NASA website:
www.nasa.gov
The US space agency hosts a massive website devoted to its spacecraft, as well as those produced in partnership with NASA. See also the NASA History homepage that includes many open-source electronic books about the history of the space age and valuable documentary materials too. Each of the NASA research centers (Ames, Armstrong, Glenn, Goddard, Johnson, JPL, Kennedy, Langley, Marshall, Stennis, and Wallops) also posts its own informative websites.

2. ESA website:
www.esa.int
The European Space Agency likewise maintains a huge online presence that illuminates its achievements in rocketry and robotic flight.

3. The Russian Space Web:
www.russianspaceweb.com
Created by space historian Anatoly Zak, the Russian Space Web is a private venture that presents reliable and in-depth accounts of Russian space projects, spacecraft, personalities, and missions.

4. Academic Institutes:
www.stsci.edu, www.swri.org
The Space Telescope Science Institute, located in Baltimore, Maryland, represents academic institutions pursuing space-based astronomical research. Its website covers the institute's operations work for the Hubble Space Telescope, the upcoming James Webb Space Telescope, and others. Similarly, the Southwest Research Institute in San Antonio, Texas, specializes (among other fields) in space exploration, which is reflected in its web pages.

5. Corporate websites:
www.spacex.com, www.boeing.com, www.blueorigin.com, www.orbitalatk.com, www.ulalaunch.com
The internet sites of SpaceX, Boeing, Blue Origin, Orbital ATK, and the United Launch Alliance profile the spacecraft and rockets originating in the private sector.

6. Encyclopedia Astronautica:
www.astronautix.com
This website presents well-researched articles about spacecraft and rockets from around the world, emphasizing their origins, history, and technology.

7. Web forum:
https://forum.nasaspaceflight.com/
This website represents perhaps the world's foremost clearinghouse for information regarding manned and unmanned spaceflight.

II. Books

The spacecraft narratives and illustrations drew on the following books.

Commercial Publishers

Caprara, Giovanni. *Il libro dei voli spaziali*. Vallardi, 1984.

Gatland, Kenneth. *The Illustrated Encyclopedia of Space Technologies*. Salamander, 1981.

———. *Manned Spacecraft*. Blanford, 1967.

———. *Robot Explorers*. Blanford, 1972.

Gatland, Kenneth, and Philip Bono. *Frontiers of Space*. Blanford, 1969.

Gorn, Michael. *NASA: The Complete Illustrated History*. Merrell, 2008.

———. *Superstructures in Space: From Satellites to Space Stations—a Guide to What's Out There.* Merrell, 2008.

Hendrickx, Bart, and Bert Vis. *Energia Buran: The Soviet Space Shuttle.* Springer, 2007.

Hunley, J. D. *Preludes to U.S. Space-Launch Vehicle Technology: Goddard Rockets to Minuteman III,* vol. 1; *US Space-Launch Vehicle Technology: Viking to Space Shuttle,* vol. 2. University Press of Florida, 2008.

Jane's Space/Spaceflight Directory. Jane's, annually since 1985.

Jenkins, Dennis. *Space Shuttle: The History of the National Space* Transportation System: The First 100 Missions. Specialty, 2010.

Miller, Ron. *The Dream Machines.* Krieger, 1993. Newkirk, Dennis. *Almanac of Soviet Manned Spaceflight.* Gulf, 1990.

Rocket and Space Corporation Energia: The Legacy of S. P. Korolev. Apogee, 2001.

Thompson, Milton O., and Curtis Peebles. *Flying without Wings: NASA Lifting Bodies and the Birth of the Shuttle.* Smithsonian, 1999.

van Pelt, Michel. *Rocketing into the Future: The History and Technology of Rocket Planes.* Springer, 2012.

NASA History Series
Gorn, Michael. *Hugh L. Dryden and His Career in Aviation and Space.* [Place of Publication]: NASA, 1996.

Green, Constance, and Milton Lomask. *Vanguard: A History.* [Place of Publication]: NASA, 1970.

Hansen, James. *Enchanted Rendezvous: John C. Houbolt and the Genesis of the Lunar-Orbiter Rendezvous Concept.* [Place of Publication]: NASA, 1995.

Logsdon, John, ed. *Exploring the Unknown: Selected Documents in the History of the U.S. Civil Space Program,* 7 vols. [Place of Publication]: NASA, 1995–2008.

Mudgway, Douglas. *William H. Pickering: America's Deep Space Pioneer.* [Place of Publication]: NASA, 2007.

NASA Historical Data Books, 1958–1998, 7 vols., 1958–1998. [Place of Publication]: NASA, 1988–2009.

Portree, David. *Mir Hardware Heritage.* [Place of Publication]: NASA, 1995.

Siddiqi, Asif. *Challenge to Apollo: The Soviet Union and the Space Race, 1945–1974.* [Place of Publication]: NASA, 2000.

Swenson, Loyd, James Grimwood, and James Alexander. *This New Ocean: A History of Project Mercury.* [Place of Publication]: NASA reprint, 1998.

III. Articles

Dupas, Alain, and John Logsdon. "Was the Race to the Moon Real?" *Scientific American,* June 1994.

Siddiqi, Asif. "Soyuz Variants: A 40-Year History." *Spaceflight,* March 2003. http://faculty.fordham.edu/siddiqi/writings/p19_siddiqi_spaceflight_2003-03_soyuz_variants.pdf

Zak, Anatoly. "Did the Soviets Actually Build a Better Space Shuttle?" *Popular Mechanics,* November 2013. www.popularmechanics.com/space/rockets/a9763/did-the-soviets-actually-build-a-better-space-shuttle-16176311/

Zak, Anatoly. "Russia's Warhorse Soyuz Space Taxi Gets a Makeover." *Popular Mechanics.* July 5, 2016. www.popularmechanics.com/space/news/a21668/soyuz-russia-spacecraft-upgrade/

INDEX

Adams, Michael, 40
Agena rocket, 60–62
Aldrin, Buzz, 16, 23, 101
Allen, Paul, 164
Alma Mater Satellite (ALMASat-1), 194
Almaz military space stations. *See* Salyut
   space stations
Anders, William, 23
Anderson, Michael, 118, 198
Andropov, Yuri, 129
Antares rockets, 189–190
Apollo program, 7, 8, 19–24, 34, 69–70,
   78, 98
Apollo–Soyuz Test Project (ASTP), 37–38,
   47–50, 69, 130
Apt, Jay, 146
Ares rockets, 156, 182
Ariane rockets, 137, 150, 193–194, 216,
   219
Armstrong, Neil, 16, 23, 40, 101
Artyukhin, Yuri, 53–54
Atkov, Oleg, 135
Atlantis, 114, 118, 121, 129–130, 132, 146,
   149, 177–178
Atlantis-Mir mission, 129–130, 132
Atlas rockets, 160, 200
Aurora 7, 15

Beggs, James, 141, 170
Belyayev, Pavel, 28, 30
BepiColombo spacecraft, 211–212
Binnie, Brian, 167
Blaha, John, 132
Blue Origin, 8, 160, 190
Bluford, Guy, 118
Boeing, 24, 161, 167–168, 178
Borman, Frank, 23
Bossart, Charlie, 59
Brand, Vance, 50
Brezhnev, Leonid, 73
Brown, David, 118, 198
Buran spaceplane, 40, 126–129, 138
Bush, George W., 121, 156, 159, 182
Bykovsky, Valeri, 27

Cabana, Robert, 179
Canadian Space Agency (CSA), 8, 211
Carpenter, Scott, 15
Cassini-Huygens spacecraft, 63, 139–141
Cassini spacecraft, 141, 150
Centaur rocket, 60, 62–63, 149
Chaffee, Roger, 23, 69
Challenger, 114, 118, 137, 144, 149, 150,
   152, 170
Chandra Telescope, 200–203, 204
Chawla, Kalpana, 118, 198
Chelomei, Vladimir, 34, 75, 110, 132
Chinese National Space Administration
   (CNSA), 8, 150, 161–163, 180–181
Churyumov-Gerasimenko comet, 219
Clark, Laurel, 118, 198
Clinton, Bill, 170
Collins, Michael, 16, 23
Colombo, Giuseppe, 85, 211
Columbia, 114, 117–118, 121, 137, 156,

159, 172, 203
Compton, Arthur H., 146
Compton Gamma Ray Observatory,
   146–149
Convair, 59–60, 62, 64
Cooper, Gordon, 15, 27
Cosmic Background Explorer (COBE), 212
Crippen, Robert, 117
Crossfield, Scott, 38, 40
CST-100 (Boeing), 160–161
CubeSat spacecraft, 194
Cunningham, Walter, 69
Curiosity rover, 7, 199–200
Currie, Nancy, 179
Cygnus spacecraft, 189–190

Deep Impact spacecraft, 207–208
Delta II rocket, 198, 207
Discovery, 114, 118, 121, 132, 144, 152,
   177–178
Dobrolovsky, Georgy, 33
Dobrovolvsky, Georgy, 53
Dordain, Jean-Jacques, 216
Dragon spacecraft (SpaceX), 159–160, 184
Dryden, Hugh L., 38
Duke, Charles, 24

Eisele, Donn, 69
Eisenhower, Dwight D., 7, 54, 78, 92
Endeavour, 117, 118, 146, 174, 177–178
Energia rockets, 137–139, 163
Enterprise, 121–126
European Space Agency (ESA), 8, 63, 118,
   137, 141, 150, 152, 172, 193–194, 207,
   211, 212, 216, 219
Explorer satelites, 59, 75–76, 76, 81, 94,
   137

Faget, Maxime, 59, 112
Falcon rockets (SpaceX), 159, 182–189
Feoktistov, Konstantin, 28
Fletcher, James, 118
Foale, Michael, 132
Fullerton, Gordon, 125–126

Gagarin, Yuri, 12, 15, 19, 26
Galileo spacecraft, 149–150
Garradd comet, 208
Garver, Lori, 182
Gehman, Harold, 121
Gemini-Agena Target Vehicle (GATV),
   60–61
Gemini capsules, 15–17, 19
Gemini-Titan II rocket, 63–66
Gidzenko, Yuri, 178
Gilruth, Robert, 15
Giotto spacecraft, 150–152
Glazkov, Yuri, 54
Glenn, John H., 12, 15
Glushko, Valentin, 126
Goddard, Robert H., 24
Goldin, Daniel, 130, 170, 172, 200
Gorbatko, Viktor, 54
Grechko, Georgy, 54
Grigg-Skjellerup comet, 152

Grissom, Virgil "Gus", 12, 23, 69
Gubarev, Aleksei, 54

Hagen, John, 56, 78
Haise, Fred, 125–126
Halley's Comet, 150, 152
Hartley 2 comet, 208
Hawley, Steven, 144
Hayakutake comet, 152
Herschel Space Observatory, 212–216
Houbolt, John C., 20
Hubble Space Telescope (HST), 141–146,
   204
Husband, Rick, 118, 198

International Space Station (ISS), 8, 47, 50,
   73, 110, 111, 114, 139, 156, 159, 163,
   168–180, 184–185, 189–190
Columbus laboratory, 174
Destiny, 173, 178
Harmony (Node 2) module, 173–174
Kibo laboratory, 174
Leonardo MPLM, 80, 174, 178
Raffaello MPLM, 180
Tranquility (Node 3), 174
Unity, 132, 173, 174–175
Zarya, 132, 173, 175–176
Zvezda, 173, 176–178
ISON comet, 208
Italian Space Agency (ASI), 139, 174, 178,
   194

James Webb Telescope, 146, 208–211
Japanese Aerospace Agency, 8, 174
Jarvis, Gregory, 118
Juno 1 rocket, 59
Juno spacecraft, 150, 204–207
Jupiter exploration, 86, 88, 91, 149–150,
   204, 207, 208

Kármán, Theodore von, 197
Keldysh, Mstislav, 75
Kemurdzhian, Alexander, 43
Kennedy, John F., 8, 15, 19, 23–24, 47, 208
Khrushchev, Nikita, 34, 75
Kizim, Leonid, 135
Klimuk, Pyotr, 54
Kohoutek (comet), 47
Komarov, Vladimir, 28
Korolev, Sergei P., 24, 26–28, 30, 34–35,
   53, 70, 73, 91, 93–94, 97, 138
Krikalev, Sergei, 132, 178
Kubasov, Valeri, 50
Kuiper Belt, 208

Landers LK lunar module, 42–43
Lazarev, Vasili, 37, 42
Leonov, Alexey, 28, 30, 50
Linenger, Jerry, 132
LISA Pathfinder, 216–219
Long March-1/2/2F rocket, 161–162,
   194–197
Lovell, James, 16, 23
Low, George, 112
Lucid, Shannon, 132

Luna spacecraft, 43, 97–101
lunar landings, 23–24, 78, 81. *See also*
    Apollo program
Lunar Orbiter satellites, 78–81
Lunokhod Rovers (1 and 2), 43–44, 101

Makarov, Oleg, 37
Malina, Frank, 197
Mariner spacecraft, 63, 82–85, 101, 211
Mars exploration, 82, 85, 86, 88, 101–102,
    197–200
Mars Pathfinder, 197
McAuliffe, Sharon Christa, 118
McCool, William, 118, 198
McDivitt, James, 16
McDonnell Douglas, 12
McKay, Jack, 40
McNair, Ronald, 118
Melville, Michael, 167
Mercury-Atlas rocket, 59–60
Mercury capsule, 12–15
Mercury exploration, 82, 85, 211–212
MESSENGER spacecraft, 211–212
MiG 105-11, 40–42
Mir space station, 8, 38, 73, 110–111, 129,
    129–130, 132, 150
Mishin, Vasily, 34, 50, 73, 132
Mojave Aerospace Ventures, 164, 167
moon exploration, 23, 43–44, 78, 81–82,
    97–98, 101
Mueller, George, 112

N1 rocket, 50, 73, 138
National Advisory Committee for
    Aeronautics (NACA), 15, 38, 59. *See
    also* National Aeronautics and Space
    Administration (NASA)
National Aeronautics and Space
    Administration (NASA), 8, 12, 19,
    37–38, 40, 60, 62, 93, 161
Neptune exploration, 86, 88, 91, 149
New Glenn spacecraft, 190
New Horizons spacecraft, 207, 208
New Shepard rockets, 190–193
Nikolayev, Andrian, 27
Nixon, Richard, 47, 88, 112, 126
North American Aviation, 20, 69

Obama, Barack, 158, 160, 174, 182, 190
Oberth, Hermann, 24
Onizuka, Ellison, 118
Opportunity rover, 44, 198–199
Orbital ATK, 8, 189–190
Orbital Sciences, 160
Orion capsule, 156–159

Patsayev, Viktor, 33, 53
Pioneer spacecraft, 63, 85–86
Planck Space Observatory, 212–216
Pluto exploration, 208
Polyakov, Valeri, 129
Polyus military spacecraft, 138
Popov, Leonid, 135
Popovich, Pavel, 27, 53–54
Progress 7K-TG, 38, 132

Proton rocket. *See* UR-500 Proton rocket

R-7 (Semyorka) rocket, 26, 70–73, 92, 94
Ramon, Ilan, 118, 198
Ranger spacecraft, 81, 97
Reagan, Ronald, 110, 118, 129, 168, 170
Redstone-Juno 1 rocket, 59
Resnik, Judith, 118
Ride, Sally K., 118
Rogers, William P., 118
Roscosmos, 7–8, 26, 47, 170, 172. *See also*
    Russian Federal Space Agency; Special
    Design Bureau 1 (OKB-1, USSR)
Rosen, Milton, 56
Rosetta spacecraft, 207, 219
Ross, Jerry, 146
Russian Federal Space Agency, 8, 137–
    138, 150, 170, 179
Rutan, Burt L., 164
Ryumin, Valery, 135

Salyut space stations, 8, 33, 37, 37–38, 44,
    50–54, 53, 73, 111, 129–130, 132, 135
Saturn 1/1B rocket, 47, 48, 66, 66–69, 69
Saturn exploration, 86, 88, 91, 139–141,
    149
Saturn V rocket, 20, 23, 44, 47, 66, 69–70,
    141
Scaled Composites, 164
Schirra, Walter, 15, 69
Schriever, Bernard A., 64
Scobee, Richard, 118
Scott, David, 16
Sevastyanov, Vitaly, 54
Shatalov, Vladimir, 33
Shenzhou capsules, 180–181
Shepard, Alan, 12, 15, 26, 40
Shepherd, William, 178
Shoemaker-Levy comet, 149
shuttle orbiters, 114–121, 121–126, 156,
    159, 163, 168
shuttle rocketry, 135–137
Skylab space station, 44–47, 50, 69, 70
Slayton, Deke, 50, 125
Smith, Michael, 118
Sojourner rover, 44, 197–198
Solar Heliospheric Observatory, 63
solar investigation, 63, 86
solid rocket boosters (SRBs), 135
Solovyov, Vladimir, 135
Soyuz 7K-LOK, 34–37, 42, 73
Soyuz 7K-OK, 30–33, 38, 106, 110
Soyuz 7K-OKS, 33–34, 53
Soyuz 7K-S, 106, 110
Soyuz 7K-ST (Soyuz-T), 106–110
Soyuz 7K-T, 37, 53, 54
Soyuz 7K-TM (ASTP), 37–38
Soyuz MS (7K-MS), 163–164
Soyuz-TM (7K-STM), 110, 129
Soyuz TMA-M, 163
Space Exploration Technologies
    Corporation (SpaceX), 8, 159, 161, 184
space launch system (SLS), 158, 182
space shuttles. *See* shuttle orbiters

Space Transportation System (STS),
    111–114, 137
Spaceship Company, 167
SpaceShipOne/SpaceShipTwo, 164–167
Special Design Bureau 1 (OKB-1, USSR),
    24, 28, 30, 33–34, 38, 73, 132
Spirit rover, 198–199
Spitzer Space Telescope, 203–204
Sputnik satellites, 56, 59, 73, 75, 91–97
Stafford, Tom, 50
Surveyor spacecraft, 63, 81, 81–82

Tempel 1 comet, 207
Tereshkova, Valentina, 27
Thagard, Norman, 132
Thomas, Andy, 132
Tiangong space stations, 180–181
Tikhonravov, Mikhail, 70, 91
Titan-Centaur rocket, 63, 91
Titov, Gherman, 27, 42
Titov, Vladimir, 132
TKS spacecraft, 110–111, 179
Tsien, H. S., 197
Tsiolkovsky, Konstantin, 24

Ulysses spacecraft, 152
UR-500 Proton rocket, 34, 50, 53, 73–75,
    101, 130
Uranus exploration, 88, 91, 149

Van Allen, James, 76, 78
Vanguard 1 rocket, 54–59, 75–76
Vanguard 1 satellite, 76–78, 94
VEGA rocket, 194
Venera spacecraft, 101–103
Venus exploration, 82, 85, 101–102
Viking rocket, 8, 75
Viking spacecraft, 63, 86–88, 197
Virgin Galactic, 167
VKK Buran spaceplane. *See* Buran
    spaceplane
Volkov, Vladislav, 33, 53
Volynov, Boris, 54
von Braun, Wernher, 19–20, 59, 66, 168
Voskhod 3KV/3KD, 27–30
Vostok spacecraft, 7, 12, 24–27
Voyager spacecraft, 8, 63, 88–91

Webb, James, 20, 70, 208
White, Edward, 16, 23, 69
White Knight spaceplane, 167
Wilkinson Microwave Anisotropy Probe,
    212
Wolf, David, 132

X-15 spaceplane, 38, 40, 112, 164
X-37B spaceplane, 167–168

Yang Liwei, 162
Yeager, Chuck, 38
Yegorov, Boris, 28
Young, John, 16, 24, 117

Zholobov, Vitaly, 54
Zuma spacecraft, 189

ACKNOWLEDGMENTS

Writing books takes lots of work. Authors certainly do their share, but without collaborators at publishing houses, no manuscript would cross the Maginot line that separates a sheaf of paper from a finished book. In the case of Spacecraft, four people at Quarto Publishing Group USA contributed immeasurably. First, Publisher Erik Gilg approved the project and set it in motion. But before that point, Senior Acquisitions Editor Dennis Pernu worked wonders guiding us through the acquisition thickets. Once we delivered our text and illustrations, Project Manager Jordan Wiklund edited the manuscript, judiciously shaping and pruning the one hundred profiles, while Art Director Cindy Laun created the stunning cover and oversaw the book's artworks. Finally, Simon Larkin did a masterful job combining the images and text in a fluid and imaginative page layout and design.

Giuseppe De Chiara

In addition to the Quarto team, I want to thank several people. First of all, I want to acknowledge the memory of my father, Antonio, who always pushed me to pursue my dreams, despite the difficulties. Also, I want to remember the late Professor Luigi "Gino" Pascale, my mentor at the University of Naples, a true master of aircraft design; Professor Francesco Saverio Marulo, under whom I completed my graduation thesis; the memory of the late Steve Pace, who introduced me to the world of professional aerospace artworks; and my friend Dwayne Allen Day at the US National Academy of Sciences. Most of all, I want to thank my wife, Annamaria Terlati, the love of my life, for her patience (since I used my few spare hours to draw the illustrations), as well my daughter, Nicole De Chiara, who is my first and best supporter.

Finally, I would like to thank my friend Michael Gorn. Without his fundamental contribution and skills, this book would not be possible. His wonderful profiles provide the essential historical and technical background to my artwork. Michael's humanity and generosity doubled the value of this project for me.

Michael Gorn

I also wish to offer a personal thank you to several individuals. I am deeply indebted to my mentor at the University of Southern California, the late historian John A. Schutz, who taught his students that effective narrative history requires good scholarship, thoughtful analysis, but above all an engaging and fluid writing style. I hope I haven't disappointed him. Memories of my friend and fellow historian Robert F. Phillips remind me that although producing historical works can be challenging, connecting audiences with the past can also be a joyous experience. On the other hand, when I groused at times about this project, I could always count on my friends Thomas and Sharon Welz for understanding and support. Most of all, I want to thank my wife, Christine M. Gorn—my guide, counselor, and soul mate—for her exceptional patience throughout the process.

Finally, I must acknowledge my collaborator, Giuseppe De Chiara. He not only produced the extraordinary illustrations that populate this book, but scrutinized the text and offered many constructive suggestions. Moreover, even though he conceived and developed the concept for this book well before I became involved, he welcomed me as a full partner. In all honesty, the best part of participating in *Spacecraft* has been the opportunity to work with this uncommonly able, generous, and open-hearted person. I am truly grateful for his friendship.

DEDICATIONS

For Christine, my beloved wife, who has the courage to be herself. —Michael Gorn

To my wife, Annamaria, and my daughter, Nicole, for their love and understanding. —Giuseppe De Chiara